T0253679

Beating the
Odds

The Life and Times of E. A. Milne

Beating the

Odds

The Life and Times of E. A. Milne

Meg Weston Smith

Foreword by Roger Penrose

Imperial College Press

ICP

Published by

Imperial College Press
57 Shelton Street
Covent Garden
London WC2H 9HE

Distributed by

World Scientific Publishing Co. Pte. Ltd.
5 Toh Tuck Link, Singapore 596224
USA office: 27 Warren Street, Suite 401-402, Hackensack, NJ 07601
UK office: 57 Shelton Street, Covent Garden, London WC2H 9HE

British Library Cataloguing-in-Publication Data
A catalogue record for this book is available from the British Library.

BEATING THE ODDS
The Life and Times of E A Milne

Copyright © 2013 by Imperial College Press

All rights reserved. This book, or parts thereof, may not be reproduced in any form or by any means, electronic or mechanical, including photocopying, recording or any information storage and retrieval system now known or to be invented, without written permission from the Publisher.

For photocopying of material in this volume, please pay a copying fee through the Copyright Clearance Center, Inc., 222 Rosewood Drive, Danvers, MA 01923, USA. In this case permission to photocopy is not required from the publisher.

ISBN 978-1-84816-907-4 (pbk)

Typeset by Stallion Press
Email: enquiries@stallionpress.com

Printed by Fulsland Offset Printing (S) Pte Ltd Singapore

For my grandchildren, Cecily, Florence, Ida, Lucas, Oscar,
Pepita, Solène, Tanguy and Thibault

A traveller who refuses to pass over a bridge until he has personally tested the soundness of every part of it is not likely to go far; something must be risked, even in mathematics.

Horace Lamb

Foreword

Edward Arthur Milne was, without doubt, a highly distinguished mathematician, astrophysicist, cosmologist and ballistics expert. From January 1929 until his death in September 1950, he held the Rouse Ball Chair of Mathematics at the University of Oxford. He was the first to occupy this chair, and (from 1973 to 1998) I was myself greatly privileged to become the third. I had known of Milne's name since my Cambridge graduate-student days, through some of his books and, most particularly, through his powerful original ideas concerning cosmology.

I became acquainted with his daughter Meg Weston Smith, the author of this remarkable biography, through a series of twenty 'Milne Lectures', held annually from 1977 until 1996 at Oxford's Mathematical Institute, these having been instigated by Meg's daughter Miranda, and generously supported by IBM (UK). The speakers were people of rare distinction and tended to have had connections with Milne in one way or another. Write-ups of these lectures were collected in the volume *The Universe Unfolding*, edited by Sir Hermann Bondi and Miranda Weston-Smith.[1] Milne's very considerable status as a scientist and mathematician was evident from the great respect afforded to him by the contributors.

It is clear that Milne's foundational work on stellar atmospheres has been deeply appreciated, and has well withstood the test of time. This is particularly impressive in view of the fact that much of this was achieved at a time when a proper understanding of the relevant quantum processes would not have been available to Milne. He made important contributions to stellar structure, and his work on ballistics was of significant importance in both world wars, his name being remembered in the basic Milne–DeMarre equation. His books and papers are respected as models of

---

[1] Bondi, H. and Weston-Smith, M. (eds) *The Universe Unfolding*, Oxford: Clarendon Press, 1998.

clarity and erudition. But for me, it was his much more controversial set of ideas in *cosmology* that have proved to be especially inspiring.

This is not to say that I believe that any version of Milne's specific cosmological model is likely to be 'correct' — it must be noted that the same comment would apply also to the models of almost *all* of the other leading cosmologists, certainly including Einstein himself, and my lack of full conviction in explicit cosmological models extends even to the popular models of today. It should be said that all these early models were basically 'guesses' of one kind or another, and this is as true of Einstein's specifically suggested models as it is of Milne's. However, Einstein's key contribution to cosmology came not from his specific proposals for a model but from his introduction, in 1916, of the *general theory of relativity* and from his realisation that this theory could have profound implications for cosmology.

Although Milne clearly recognised the revolutionary power of Einstein's general theory, he appears to have been reluctant to take it fully on board. This could well have contributed also to the undoubted limitations of his otherwise well-thought-through picture of stellar structure, these limitations becoming relevant when it came to the study of the central regions of particularly massive stars. Yet Milne was far from alone in this lack of conviction concerning the implications of general relativity for the structure of such stars: others, including the great astrophysicist Sir Arthur Stanley Eddington, one of the staunchest supporters of Einstein's theory — and even Einstein himself — did not fully appreciate some of the extraordinary implications of general relativity, in this context, these leading to our present viewpoint concerning *black holes.*

It was not at all unreasonable, therefore, that at the time when Milne was developing his cosmological ideas, schemes other than those that were strictly in accordance with general relativity should be presented as serious alternatives. Milne's own distinctive approach to cosmology, which he referred to as *kinematic relativity*, paid scant attention to Einstein's motivations for regarding gravitation as a feature of space–time curvature, and was concerned basically with the space–time geometrical aspects of a spatially uniform expanding Universe. Although Milne's theory ignored the revolutionary insights that Einstein had introduced concerning gravitation, it was by no means simply a retreat to the older

Newtonian ideas of space and time, providing instead a novel and far-reaching outlook concerning the basis of space–time geometry. One aspect of this was to concentrate on *time* rather than space — the notion of separation that is ascertained by *clocks*, rather than by rulers — as being the basic measure of space–time geometry. Light signals provided the link that enabled space measures to be derived from time measures, where spatial distances were to be derived through the use of *radar*-type determinations. This very insightful idea of basing space–time geometry on time measurements rather than spatial ones (which had, in essence, also been previously put forward by the leading Irish relativity theorist John Lighton Synge in a brief paper published in *Nature* in 1921,[2] and often promoted by him later) was, at the time, met with much scepticism, and even derision. However, Milne's (and Synge's) perspective has now proved to be victorious, even to the extent that nowadays the very definition of the *metre* is given in terms of a time, namely as *exactly* 1/299792458 of a light second!

The metre rule in Paris is now much too poor a standard to be used for the precision needed for modern physical measurements, whereas time measures are far more precisely determined. The basic physical reason for this time-measurement precision is that individual (stable) massive quantum particles are, in a clear sense, themselves superb clocks, 'ticking' away with a frequency ν that is precisely determined by the particle's mass m — via the basic physical constants h (Planck's constant) and c (the speed of light). This comes about from combining Einstein's famous $E = mc^2$ with Planck's $E = h\nu$, which tells us that the 'tick rate' of a stable particle is simply proportional to its rest mass m: $\nu = m \times (c^2/h)$. Although *single* particles themselves do not make usable clocks, the very great accuracy of modern atomic and nuclear clocks rests ultimately on this very basic principle.

Rulers, on the other hand, provide relatively poor measures of distance. Quite apart from their being subject to all sorts of extraneous effects such as temperature and gravitational influences — and depending, as they do, on the notion of *rigidity*, a concept that is somewhat foreign to relativity theory — rulers are also *geometrically* inappropriate, in space–time terms,

---

[2] Synge, J.L. 'A system of space–time coordinates' in *Nature* 108 (27 October 1921), p. 275.

because a ruler has a history that describes a two-dimensional *strip*, which cannot measure separation between two events (space–time 'points') unless these two events are *simultaneous* in the ruler's rest-frame. To ascertain the required simultaneity, we would need Einstein's procedure using light signals, or something equivalent, which is much more complicated than the simple use of a clock. In any case, we still need the light-rays, so we are basically back to Milne's radar prescriptions. On the other hand, the geometrical space–time description of a small *clock*'s history — or *world-line* — is a one-dimensional *curve*, and the space–time 'length' of that curve, separating two of its events, is simply the time elapsed between those events as measured by the clock, and no awkward issues of 'simultaneity' arise.

Once we have our accurate, physically well-defined clocks, in order to determine temporal separations between events, we can use Milne's 'radar method' to define spatial separations, and the space–time geometry is thereby obtained. Milne's method of defining the distance between space–time events in terms of reflected light signals requires that the speed of propagation of light signals is embedded in the space–time geometry, i.e. we need to know the space–time curves representing the possible world-lines of *photons*. These are space–time curves along which the time measure would be *zero*. The space–time information that determines these world-lines is called the 'light-cone' structure, or in more technical geometric language, the *conformal* structure of space–time. The full space–time structure is fixed once we know the conformal structure, together with the time measure provided by ideal clocks.

Milne's 'clock and light-ray' approach to looking at space–time geometry is not merely pedantry, as it enables us to move beyond the restrictions of standard general relativity. It proves to be extremely valuable to *separate* the 'rate of time's passage' from the conformal ('light-ray') structure, which demonstrates the prescient nature of Milne's (and Synge's) insights. Indeed, James Clerk Maxwell's wonderful fundamental system of equations for electromagnetism, formulated in 1865,[3] does *not* depend on the 'clock' aspect of space–time structure and needs only the

---

[3]Maxwell, J.C. 'A dynamical theory of the electromagnetic field' in *Philos. Trans. Roy. Soc. (Lond.)* 155 (1865), pp. 459–512.

conformal structure. The behaviour of massless particles and fields (gravity excepted) needs only the conformal structure, and does not require the notion of the rate of time's passage, as measured by clocks. Accordingly, a significant part of physics — namely that for which the notion of *mass* plays no role — needs only the *conformal* part of the space–time metric, and this allows the notion of clock rates to be separated from the other aspects of space–time geometry.

This general way of thinking led Milne to his most revolutionary and apparently outrageous idea: that it is not only *possible* to separate the measure of time from the geometry of light propagation, but that one might seriously envisage the possibility of *two* quite distinct notions of time somehow co-existing within one Universe!

In order to understand Milne's motivation for introducing such a strange idea, we should bear in mind that at that time there was an alarming discrepancy between the estimated age of the Universe, from the big bang to the present time, and the estimated age of some very ancient collections of stars called *globular clusters* — some hundreds of light years across and containing thousands of stars. The globular clusters appeared to be considerably older than the Universe itself, which was a serious contradiction! It was this contradiction that would later give rise to the *steady-state model* of the Universe of Hermann Bondi (later Sir), Tommy Gold and Fred Hoyle (later Sir), in which the eternally expanding Universe was envisaged as having its material replenished by the continual creation of hydrogen molecules throughout space, so the total time scale of the Universe could be infinite, and thus not contradict the ages of globular clusters. However, Milne had earlier chosen an even more ingenious route for resolving this contradiction. He noticed that the estimated age of globular clusters depended on a measure of time that was obtained through dynamical (gravitational) processes — which he referred to as *dynamical time* — whereas the age of the Universe had been determined via its expansion rate, which required good estimates of the distances of other galaxies, these in turn depending upon knowing the intrinsic luminosities of certain individual stars. The stars in question were those known as 'Cephied variables', whose periods of variation could be precisely related to their luminosities, through the theoretical understandings of stellar structure that

Milne himself had been seriously involved with. The measure of time that Milne was led to consider for determining the age of the Universe was what he called *atomic time*, and he noticed that the assumed equality between these two measures of time was an unappreciated *assumption* — which he considered might well be false!

Accordingly, Milne was able to provide a resolution of this cosmological paradox by assuming that clocks that were constructed to measure dynamical time would have run much more rapidly in the past, in comparison with those constructed to measure atomic time. In this way he was able to provide a theory in which the Universe extended indefinitely into the past, as measured by the dynamical 'clock' provided by the globular clusters according to which there would be no 'beginning' to the Universe, while according to the 'atomic time' that was ascertained from the Cephied variable stars in distant galaxies, the Universe would have had an origin in the finite past.

As things turned out, the resolution of the paradox concerning the age of the Universe came from a different direction from either that of the steady-state model of Bondi, Gold and Hoyle or the 'two-times' resolution previously provided by Milne. It eventually turned out (mainly through the work of Walter Baade and Allan Sandage in the 1950s) that the age of the Universe, as determined by the Cepheid variables, was badly underestimated by an overall factor of nearly three. This was largely owing to a confusion between two *different* kinds of Cepheid variables that turned out to exist, and the discrepancy with the age measured by the globular clusters disappeared.

Nevertheless, Milne's idea of the possibility of more than one different measure of time, consistently with the same conformal structure, had made its mark on me. I even recall reading J.B.S. Haldane's extraordinary essay[4] in which he envisaged the origin of life coming about in accordance with Milne's vision of the progressions of time and physical constants. Although, like so many other proposed models of the overall history of the Universe, Milne's particular scheme is not now taken seriously, his insights into the nature of time have opened our eyes to other possibilities.

---

[4]Haldane, J.B.S. 'Mathematics and Cosmology', in *The Marxist Philosophy and the Sciences*, London: Allen & Unwin, 1938, pp. 66–73.

Indeed, at the time of writing, I am strongly of the opinion that such insights will play an important role in a revolutionary new outlook on cosmology that may shortly be forced upon us. However this works out, our debt to Milne's cosmological insights is undoubtedly profound.

Roger Penrose

Preface

My biography of my father, who died when I was a schoolgirl, is born of curiosity. In 1945, on returning home to Oxford after wartime evacuation in America, I found him frail and elderly although he was not yet fifty. Ravaged by Parkinsonism, he had a palsied arm, a glazed facial expression devoid of emotion and a disinclination to speak that created a barrier to communication. Nobody thought to explain that he was suffering the delayed effects of an illness he had as a young man, nor to enlighten me about the deaths of my mother and step-mother — a taboo subject.

The annual Milne lectures, from 1977 to 1996, and an album of newspaper cuttings inspired me to search for a person I had not known and my quest has been highly rewarding, scientifically and personally. To my amazement, his friends and colleagues spoke of my father's vivacious conversation, his ferocious energy and his unusually fertile mind. His bold, unorthodox, sometimes extravagant ideas often provoked controversy and drove new thinking. From his letters, in European and American libraries and in private hands, I glimpsed his innermost feelings and more than once I have been touched to be sent a single handwritten letter preserved for many years.

My father grew up in a different scientific milieu, without space travel, and was already in his prime when the minor planet Pluto was discovered and when the knowledge of the expansion of the Universe altered world perception. Beyond cosmology and astrophysics, he made formidable contributions to defence. His pioneering ballistics for World War I were the springboard for his accelerated rise in science. By the age of thirty he was a professor and Fellow of the Royal Society although he had not completed an undergraduate degree. Piecing together his work in World War II from tissue-thin documents at the National Archives was a fascinating exercise.

"Tell us about the man; don't worry about the science", advised the Nobel laureate S. Chandrasekhar, and I have endeavoured to show the human face of science by charting my father's struggles and triumphs in the context of the people and circumstances that shaped the course of his life and his outlook. He overcame personal misfortunes of financial hardship, debilitating illness and devastating bereavement with gritty tenacity strengthened by deep-seated Christianity. Obsessed by the superior power of mathematics over physical reality, brilliant yet gauche, he emerges as a scientific heretic.

Acknowledgements

I could not have pieced together my father's life without the universal co-operation of a vast number of people, especially Dr Subrahmanyan Chandrasekhar, Lady Jeffreys, Sir William McCrea, Professor Geoffrey Walker and Professor Gerald Whitrow.

I am obliged to Commander Witts RN for taking me round HMS *Excellent*, Whale Island; to staff at the Shuttleworth Collection, Bigglewade, for giving me privileged access to its early planes; to Dr Robin Catchpole and Dr David Dewhirst of the Cambridge Observatories; and to Professor Ernest Walton for retracing with me my father's last footsteps in Dublin.

I am grateful to Dr John Milne and Dr Christopher Coulson for family history; to Brian Bower, Harry Coulson, Francis Scott and Harry Wright for recalling my father's schooldays in Hull; to Sir John and William Barnes for information about their father Bishop E.W. Barnes; to Edwin Norris and George Tyson for telling me of their Manchester days; to John Bamborough, Ian Crombie and Sir William Deakin for their memories of Wadham College; to Dr Roger Hutchins for summarising the Radcliffe Observatory saga; and to Dr Madge Adam, Giles Barber, Dr Kenneth McNeill, Dr Kevin Westfold and Dr Richard Wilson for illuminating Milne's Oxford years. His graduate students were unfailingly forthcoming: Dr Leslie Camm, Professor Gabriel Cillié, Professor Tom Cowling, Ivan A. Getting, Professor Thornton Page, Dr Chandrika Prasad and Dr Geoffrey Wiles.

My warmest thanks to all who encouraged me by responding to my questions, giving me introductions, lending me carefully preserved letters and reading parts of my manuscript: Professor Maurice Bartlett, Professor Nick Bingham, Professor Brevis Bleaney, Professor Gustav Born, Dr Jean Bradley, Dr Stephen Bragg, Dr Hermann Brück, Professor John Burkill, Betty Burn, Dame Mary Cartwright, Cecil Chapman, Jean Cohen, John Collie, Kenneth Coulson, Sir David Cox, Professor Donald Coxeter, Professor Richard Crane, Helen Cranswick, Michael Crum, Allan Curtis, Mary Cuthbert, Anne Davenport, Dr Tom David, Cliff Davies, Jeffrey

Dean, Dr David DeVorkin, Professor Michael Dewar, Professor Arthur Dickens, Professor Paul Dirac, Professor Patrick Duff, Miriam Dunham, Hugh Elsey-Warren, Professor David Evans, Marjorie Farrer, J.R. Fewlass, Felicity Foster Carter, Win Fryer, Professor George Gale, Dr Bob Gifkins, Professor O. Godart, Dr George Gordon, Professor Walter Gratzer, Professor Jeremy Gray, Dr Victor Hale, Julia Hanford, Katherine Haramundanis, Elaine Hartree, Oliver Hartree, Professor Agnes Herzberg, Professor Mary Hesse, Dr Norriss Hetherington, Jean Hewitt, Greta Hopkinson, Karl Hufbauer, Dr Christopher Jeans, Professor Eberhart Jensen, Elsie Johnson, Professor Reginald Jones, Jet Katgert, Professor David Kendall, Professor Helge Kragh, Dr Gary Krenz, Dr Sheila Lee, Dr Elizabeth Leedham-Green, Dr John Lennard-Jones, Professor Per B. Lilje, Sir John Maddox, Kathleen Major, Brigadier Andrew Mayes, Professor Hugh Michael, Alan Milne, Kathleen Milne, Mollie Milne, Professor Philip Moon, Anthony Mulgan, Dr Vicky Murphy, Tressilian Nicholas, Professor Igor Novikov, Professor Jan Oort, Jonathan Parkin, Sir Henry Phelps Brown, Professor A. Phillips Griffiths, Barbara Pidgeon, Sir Brian Pippard, Mary Plant, John Plaskett, Sir Karl Popper, Dr Dorothy Rayner, Dr Margaret Rayner, Simon Rebsdorf, Professor Graham Richards, Basil Rimes, Reverend Stephen Roberts, Dr Robert Robson, Professor Ian Roxborough, Dr John Sanders, Mary Scott, Professor Alexandr Sharov, Professor David Shoenberg, Professor David Singmaster, Professor Cedric Smith, Dr Harlan Smith, Dr Frank Smithies, Dr Seymour Spencer, Professor Walter Stibbs, Walter Strachan, Dr John Synge, Dr Derek Taunt, Dr Paul Taylor, Dom George Temple, Mary Thompson, Geoffrey Turberville, Professor Albrecht Unsöld, Professor Kamesh Wali, Professor Joan Walsh, Sylvia, Countess de la Warr, Captain John Wells RN, Dr Adriaan Wesselink, Magda Whitrow, Professor Ian Williams and Professor Jens Zorn.

Alan Kucia at Trinity College patiently answered my enquiries, as did Peter Hingley at the Royal Astronomical Society. I received generous help from officers and staff of the Cambridge Philosophical Society, University of Durham Philosophical Society, Hull Literary and Philosophical Society, Leeds Astronomical Society, London Mathematical Society, Manchester Literary and Philosophical Society, Oxford University Natural Science Club, the Royal Institution, the Royal Meteorological Society, the Royal

Society of Edinburgh, the Royal Society and the Trinity Mathematical Society.

I am obliged to archivists and librarians: in Cambridge at Caius College, Churchill College, Clare College, King's College, Newnham College, Peterhouse, St John's College Scientific Periodicals Library, the Sidney Sussex College, Trinity College, the University Library and the Whipple Museum; in London at the British Library, Imperial College, the Imperial War Museum, the Institute of Physics, University College, the Wellcome Institute and the National Archives at Kew; in Oxford at the Bodleian Libraries, Christchurch, the Clarendon Laboratory, Hertford College, the Museum for the History of Science, New College, Nuffield College, Queen's College, the Radcliffe Science Library, Rhodes House, St Hilda's College and Wadham College.

Also my thanks to those at Eton College; Hymers College; Withington Girls' School; the Royal Aircraft Establishment, Farnborough; the University of Glasgow; Central Library, Halifax; Hendon Air Museum; Humberside and Brynmor Jones Libraries in Hull; Brotherton Library, Leeds; Central and John Rylands Libraries in Manchester; the Ministry of Defence; Central Library, Portsmouth; and the University of Sussex. In the United States my thanks go to staff at the American Institute of Physics; the Bentley Historical Library, Ann Arbor, Michigan; the California Institute of Technology; the University of Chicago; Harvard University; the Huntington Library; the University of Illinois, Urbana; Princeton University; and the University of Texas at Austin. Thanks also to staff at the Hebrew University of Jerusalem, Israel; the Catholic University of Louvain; the Deutsches Museum, Munich; the University of Oslo; and the Nehru Memorial Library, New Delhi.

I wish to record particular gratitude to Professor Sir Roger Penrose, who held the same Oxford chair as my father, for kindly writing the Foreword. I am also indebted to Professor Ernst Sondheimer for scrutinising the entire manuscript, and improving the text with corrections and constructive comments, as did Professor Leon Mestel and Professor Donald Lynden-Bell. I owe much to Professor Virginia Trimble.

My warm thanks to my late husband John, my son-in-law Professor Matthew Bell and especially my daughter Miranda Barlow for their invaluable and steadfast support.

I wish to make grateful acknowledgement to all copyright holders and to the following for their kind permission to reproduce material.

The Master and Fellows of St John's College and Trinity College, Cambridge, and Balliol College, Oxford. The Warden and Fellows of Nuffield College and Wadham College, Oxford. The Bodleian Libraries of the University of Oxford. The syndics of Cambridge University Library. Imperial College and University College London. The Royal Institution of Great Britain. The Royal Sociey. The American Institute of Physics. Harvard University Library. Princeton University Library. National Archives of Norway, Oslo. Members of the families of Sir John Cockcroft, Professor A.V. Hill, Sir Francis Pridham and Sir William McCrea.

Although I have tried to locate copyright holders I will readily correct any inadvertent errors and omissions.

List of Figures

Abbreviations in the Footnotes

AGW denotes the papers of A.G. Walker, Balliol College, Oxford.

AVHL denotes the papers of A.V. Hill, Churchill Archives Centre, Cambridge.

EAM denotes the papers of E.A. Milne, the Bodleian Libraries, Oxford.

GJW denotes the papers of G.J. Whitrow, Imperial College London.

NA denotes the National Archives, Kew.

SHPQ denotes Sources for the History of Quantum Physics.

SR denotes the papers of Svein Rosseland, National Archives of Norway, Oslo.

WHMcC denotes letters from Milne to Sir William McCrea, now lost.

Contents

Contents

Chapter 1

A Foothold on the Ladder

The extent and condition of secondary education in Hull was something of a scandal.[1]

In the early 1930s Albert Einstein[2] often came to Oxford. He was the guest of the physicist Professor Frederick Lindemann,[3] later Lord Cherwell, advisor to Winston Churchill, who was eager to encourage contact with German scientists. During one visit someone asked Einstein what most impressed him about Oxford, to which he replied 'E.A. Milne'.[4] The reason was not hard to find. At a time when the university marginalised science, my father stood out for his youthful vitality, brimming with ideas. He was thirty-two when he took up his chair in mathematics — despite having no degree. A magnet for students of the highest calibre, he energetically put new life into a moribund department, and established an international centre of astrophysics. That he scaled the heights of academia was no surprise to his cousins, who in childhood had jeered his bookish inclinations, but, given his humdrum background, his rise was remarkable and rapid. Until he was seventeen he never left Yorkshire.

Edward Arthur Milne, known as Arthur, was born in Hull on 14 February 1896, the eldest child of primary school teachers. Sidney Arthur Milne and his wife Edith (née Cockcroft) were saturated in the scriptures and the three R's, writing, reading and arithmetic, along with

---

[1] Lawson, J. 'Middle-class Education in Later Victorian Hull' in *Studies in Education, Journal of the Institute of Education, University of Hull*, 3, 1 (1958), p. 27.

[2] Albert Einstein (1887–1955). In 1914, after appointments at Zurich and Prague he became the Director of the Kaiser Wilhelm Institute of Physics, Berlin. From 1933 he had a post at the Institute for Advanced Studies, Princeton. Nobel Laureate 1921.

[3] Frederick Alexander Lindemann, Viscount Cherwell (1886–1957). Dr Lee's Professor of Physics at Oxford from 1919 to 1956.

[4] Letter from M.M. Crum to the author, 13 February 1991.

a smattering of history and geography, and they paid not the slightest attention to the revolution in physics that took place during Arthur's boyhood. A few weeks before his birth, Roentgen[5] discovered X-rays. When Arthur was in his pram, J.J. Thomson[6] identified the electron. When he was five, Marconi[7] transmitted radio waves across the Atlantic; when he was eight, Rutherford[8] and Soddy[9] formulated the laws of radioactivity; and during his teens Rutherford and Bohr[10] postulated the nuclear structure of the atom. It was fortunate for Arthur that his schoolmasters fired him with curiosity about the fundamental discoveries which heralded the atomic age.

Our Yorkshire roots went back for generations. The Milnes, God-fearing, respectable, unadventurous, had multiplied in and around Hull and my father learnt the names of his twenty-nine Milne first cousins by reciting them to a hymn tune. The vast family network of shopkeepers, nurses, teachers, accountants and clerks penetrated every crevice of the city's industry and commerce. A job to be coveted was one at Reckitt's, the philanthropic Quaker firm, because it provided sick pay and pensions long before government social welfare.

Sidney's father Joshua, a customs officer, and his wife Eliza (née Dixon) were unusual parents in that they raised all thirteen of their children to maturity, unscathed by lethal epidemics of smallpox and scarlet fever. In their small cramped house at 48 Louis Street, just off Spring

---

[5] Wilhelm Konrad von Roentgen (1845–1923). German physicist. Nobel Laureate 1901.

[6] Sir Joseph John Thomson (1856–1940). Cavendish Professor of Physics from 1896 to 1919. Master of Trinity College from 1918 to 1940. President of the Royal Society from 1915 to 1920. Nobel Laureate 1906.

[7] Guglielmo Marconi (1874–1937). Radio engineer. Nobel Laureate 1909.

[8] Ernest Rutherford, Baron Rutherford (1871–1937). Langworthy Professor at Manchester from 1907 to 1919 and Cavendish Professor at Cambridge from 1919 to 1937. Nobel Laureate 1908.

[9] Frederick Soddy (1877–1956) worked with Rutherford at McGill University on radioactive decay, foresaw the method of 'carbon-dating' and the potential use of uranium for atomic energy. At Oxford, disillusioned by its arrangements for research in chemistry, he busied himself with economic and monetary reform. Nobel Laureate 1921.

[10] Niels Henrik David Bohr (1885–1962) worked with Rutherford at Manchester before becoming head of the Institute of Theoretical Physics in Copenhagen. Nobel Laureate 1922.

Bank, the children ate their food sitting on the stairs, each one in his or her allotted place. Joshua was given to practical japes, and one morning as he left for the docks, he exhorted the family to be on their best behaviour at their midday meal as he was bringing home a stranger, a black man. They welcomed the black man to their meal and all was going nicely until one boy spotted that one of his brothers was missing, and piped up, 'Where is Fred?' End of prank; the black man was Fred.

Edith's early forebears were cordwainers (shoemakers) in Hebden Bridge, the heart of the boot industry; more recent Cockcrofts were surgeons and architects. Her maternal grandfather, George Hepworth, was the founder-builder of the village of Brighouse, and one of her Hepworth uncles was a local historian and noted photographer. Yet Edith endured a harsh childhood and she never had the chance to go to secondary school because when her mother Ann died, Edith had to stay at home to care for her four younger sisters and brothers. Her father William Cockcroft, stationmaster at Pontefract, moved the family to Hull and took a job as a cashier. He married Mary, the widow next door, and she became a much loved step-mother and step-grandmother. Released from domestic drudgery, Edith enrolled as a 'pupil-teacher' at a primary school, where she leant from her seniors while teaching younger children. The certificate she gained put her in the first division and, in turn, she supervised trainees. When she got engaged to Sidney at the age of twenty-two she was earning £55 a year.

Small, wiry and red-headed, Edith had a harder, less sunny personality than Sidney. She harboured resentments and grumbled about life's vicissitudes; neurotically preoccupied with cleanliness, she was ever scouring and scrubbing. She had a good head for figures and was easily as intelligent as Sidney, but he was better qualified. After getting his pupil-teacher certificate, he studied at the Diocesan Training College (now St John's College) in York and won a divinity prize. He was a devout Christian and his family thought he might enter the church but he regarded his work as a schoolmaster as a form of ministry.

Sidney was twenty-three when he was appointed headmaster of a Church of England primary school for boys in a rough neighbourhood. His pupils were mainly sons of dock labourers and mill workers, and occasionally boots were distributed to them. The school in Salthouse Lane, near the

Queen's Dock (now the Queen's Gardens), was wedged between a public house and a warehouse and noise penetrated its dilapidated thin walls, making even harder the task of teaching more than one class in the main room. Under these trying circumstances Sidney's job was difficult but he was firm, calm, perhaps a trifle too serious, and earned plaudits from the people at St Mary's Church in the Lowgate for his integrity and upright Christian rectitude. His pupils were in awe of him, his nephews feared him not a little and Arthur was profoundly conscious of his strong moral influence.

Sidney and Edith were steeped in the Victorian precepts of duty, loyalty, obedience, respect for authority and veneration of their elders, precepts drummed into Arthur's psyche. Above all they were obsessed with the imperative for self-improvement. Trapped in a rut by their mediocre qualifications that denied them advancement, they were adamant that their sons should escape a similar yoke. Without access to wealth or influential connections, the accepted pathways of the privileged, the passport to a better life lay in education. Their sons must attain a higher level than they had, and it was in a household deeply committed to the value of education that Arthur learnt his alphabet and numbers.

Arthur's relatives pooh-poohed his parents' lofty ambitions. In their eyes, employment took precedence over 'book-learning', which they scorned as a risky irrelevant indulgence that merely postponed the day a youngster could contribute to the family budget. And there were jobs aplenty. Arthur grew up in an imperial age of optimistic civic pride. Hull was at the height of its vibrant bustling prosperity inextricably linked to seafaring. Nine docks spread across 145 acres processed grain and coal arriving by river and canal. The smart sailor suit which Arthur wore for Sunday best proclaimed Britain's maritime supremacy. Yet life at sea was cheap, and this brought the fear of destitution. The stigma of debt was unpardonable and if the breadwinner died or became incapacitated the dreaded workhouse beckoned poverty-stricken widows and orphans.

Edith harboured social pretensions. She aspired to something better than their mean little terrace house at 4 Cromwell Street (now a small park), and when Arthur was five and his brother Geoffrey three they moved to Suffolk Street in Newland, first to number 41 then to 38. At the nearby council school in Lambert Street, Arthur had his first taste of formal learning. Newland was some two miles from St Mary's but Sidney insisted that they worship there. The long walk there and back stuck in Arthur's memory for his aching legs but he liked sitting in the pew alongside his cousins Clifford and Elsie Milne. Four years later the family moved to Northfield, an isolated cluster of a dozen houses surrounded by fields, close to the village of Hessle. Edith took pride in the small front garden which took them up a notch on the social scale, yet living at Northfield set them apart from their relatives in Hull because the journey was a major undertaking involving a horse-drawn tram and a bus. That is not to say Arthur never saw his relations. He was especially fond of his Cockcroft step-granny and her cooking. In the room she allocated to photography he watched spellbound as his Hepworth great uncles operated the enlarger.

The Milnes were not unusual in having no wireless or telephone, and outside sanitation was the norm. Horses and carts were a common sight. Men dominated nearly every walk of life. Women did not have the vote, could not practise the law or serve in the police, and once she had married, Edith was barred from teaching in a government school. Hull had no public museum or art gallery and the world beyond scarcely existed, for what was Yorkshire was best. This limited outlook was offset by the precious advantage that Arthur had — a stable and secure boyhood.

Community life revolved around the church, with picnics and parades. Uppermost among the issues of the day was the sanctity of Sunday. Should the landing of fish be allowed? Or music in the park? Sidney switched allegiance to the parish church in Hessle, All Saints, where he was a lay reader and established a men's club. Arthur grew up accustomed to grace before meals, bedtime prayers and the strict observance of Sunday, attending two services and Sunday school. Throughout the day Arthur and Geoffrey had to comport themselves with seemly sobriety, reading the scriptures or memorising passages from the Bible, especially the Psalms, certainly not playing cards. Arthur's detailed understanding of the Bible, church history and ritual, its calendar, festivals and liturgy was such that years later a vicar paid him the compliment of being 'as good as a curate'.[11]

---

[11] Attributed to Canon Frederick William Dillistone (1903–1993), vicar of St Andrew's Church, Northmoor Rd, Oxford, from 1934 to 1938. Letter from T.G. Cowling to the author, 10 December 1989.

Although Arthur attended the Church of England National School in Hessle, his real education took place at home. Sidney was good with boys and, realising that Arthur was a child of exceptional promise, assiduously nourished his intellect. Sidney's interests were primarily theological and literary but he had an immense store of general knowledge which he gently imparted during walks along the Humber estuary. Arthur admired the skill of men who caught with tongs the red-hot rivets dropped from above. At the Victoria and Minerva piers they watched the busy traffic of trawlers, keels, dredgers, cargo and passenger ships. Conscientious to a fault, Sidney borrowed library books to help Arthur improve his proficiency in chess and encouraged him to read traditional schoolboy fare like *Robinson Crusoe, Masterman Ready* and *The Swiss Family Robinson.* Through the writings of J.C. Squire, Max Beerbohm, Edward Lear and Lewis Carroll, Arthur developed a penchant for parody, clerihews, limericks and nonsense verse, which was to stand him in good stead.

Arthur and Geoffrey had different temperaments. Arthur, slight and small like his mother, inherited her passion for detail and capacity for rapid and copious talk. His bright vivid smile lit up his round face topped by light brown hair, and he wore glasses. If he owed his brains to the Cockcrofts and Hepworths, he absorbed Sidney's grit, probity and self-discipline. Geoffrey, who resembled Sidney, was bigger, sturdier and less highly-strung than Arthur (Fig. 1).

The parents drove their sons rather hard, always expecting them to be engaged in something purposeful, so that, according to their cousins, they missed out on their share of rough and tumble. The picture that emerges of Arthur is of an amiable and studious little boy, rather intense, remote and not given to quip or jest.

It would be unjust to think that Sidney and Edith neglected leisure pursuits. Although they took their sons to Hull's annual fair where the boys liked going on the hand-cranked merry-go-round and hurtling down the spiral chute of the 'lighthouse', their parents preferred more edifying activities. They joined the craze for bicycling, and on excursions into the countryside their destination would be a village church. Sidney instructed the boys on ecclesiastic architecture and on heraldry from mural tablets with coats of arms. Afterwards Sidney supervised the boys' carefully labelled drawings. These subjects sit well together and Arthur assimilated a sound knowledge of them and a love of their specialised vocabularies which he enjoyed rolling off his tongue.

Sidney sharpened Arthur's mind with the express aim of getting him into a good secondary school. But education was a scarce commodity and the situation in Hull was particularly dire. Although the national school-leaving age was officially twelve, it was perfectly legal for children aged eleven to be wholly or partially excused attendance and there were some 70,000 half-timers in the country. (Not until 1918 was full-time attendance compulsory up to the age of fourteen.)[12] The provision in Hull was disgraceful, one of the worst in the country. When Arthur was born and the population of Hull was about 200,000, only a dismal 500 boys and girls received secondary schooling from the government.[13]

In this educational desert a fee-paying school for boys had recently opened. Set in spacious grounds and with well-qualified staff, Hymers College[14] quickly gained an enviable reputation under its outstanding headmaster Charles Gore.[15] The school was expensive, certainly far too expensive for the Milnes to afford on Sidney's small income. Ambitiously they pinned their hopes on Arthur winning one of the scholarships available to boys from elementary schools. Alive to the intense competition, Sidney tirelessly prepared Arthur for the examination. He was the catalyst for his studies, guiding, stimulating and enriching the books he prescribed.

For summer holidays the Milnes stayed at boarding houses in Scarborough or the less-fashionable resorts of Hornsea and Withernsea, easily reached by train on a cheap weekly ticket. They were lodging with a Mrs Brown in Hornsea when the postman brought the anxiously awaited envelope with the examination result. It bore the joyful tidings that Arthur

---

[12] Howson, G. *A History of Mathematics Education in England*, Cambridge: Cambridge University Press, 1982, p. 277.

[13] Lawson, J. 'Middle-class Education in Later Victorian Hull', in *Studies in Education, Journal of the Institute of Education, University of Hull*, 3, 1 (1958), p. 47.

[14] The Reverend Dr John Hymers (1803–1887). Mathematician. Fellow of St John's College, Cambridge, from 1827 to 1853. He left his fortune, amassed from railway shares, to the Corporation of Hull. Hymers College opened in 1893, achieved full independence in 1976 and today is co-ed.

[15] Charles Henry Gore (1862–1945). Headmaster of Hymers College from 1893 to 1927.

had triumphed. The splendid scholarship covered all his school fees plus his train fares between Hessle and Hull. Even more importantly the scholarship secured his crucial foothold on the first rung of the educational ladder. Two years later Sidney prepared Geoffrey with equal success and the significance of their scholarships[16] cannot be overstressed for Hymers opened up possibilities far beyond the narrow concepts of their family.

As the new term approached, Arthur, now in possession of a Norfolk jacket and stiff Eton collar, was beside himself with excitement: 'I could scarcely sleep either for thinking of my locker and locker-key ... and then to my disappointment the first day ... was only a half-day and it seemed an age till the next day.'[17] Scholarship boys like Arthur skipped the lowest form of the senior school as did abler boys from the junior school, who initially looked down on the newcomers. His first year was uncomfortable. That Arthur eagerly lapped up lessons with earnest, rapt attention made him the butt of his classmates, who nicknamed him 'Swotty Blinks' for his glasses and lack of sporting ability. In class he was an accomplished all-rounder, to the extent that his peers felt they stood no chance because he swept up prizes across the board. To avoid their taunts he sometimes deliberately introduced errors into his written exercises but this did not stop his gaining an extra £10 towards his expenses at the end of his second year.

Hymers did not have catering facilities and most boys went home for lunch. Arthur took sandwiches and occasionally went to his Aunt Wonty, Sidney's younger sister. [18] Arthur's cousins liked to gloat over one of his lunchtime visits to his step-granny because it showed his abstraction from practicalities. Placing a shilling in his hand she asked him to run an errand. Before setting off, Arthur went to the privy in the garden and he must have dropped the coin in for it was never seen again but his cousins made sure it was not forgotten.

---

[16] When Hymers applied for direct grant status, it cited Arthur as an example of the benefit of scholarships. Scott, F.W., Sutton, A., King, N.J. *Hymers College*, Beverley: Highgate Publications, 1992, p. 96.

[17] Letter from Milne to the author, 23 November 1942.

[18] Wonty was an elision of 'one-too-many', a typical piece of Milne humour. Her son Cuthbert had a knack for invention, and his nutcracker, the Crackerjack, that gently broke the shell with a ratchet keeping the kernel intact, was a huge commercial success.

The boys at Hymers greatly admired their headmaster. Courteous, distinguished-looking Mr Gore believed in a balanced curriculum. At impromptu debates of the whole school he chose a speaker by putting his pencil at random on the school list. On one occasion he called on Geoffrey to propose the motion, 'I am a wise man'. Thoroughly primed in the scriptures, Geoffrey declared that he knew he was a fool before citing Solomon: He who knows he is a fool is a wise man.

The debates took place in the central hall surrounded by classrooms that were dimly lit and chilly because the smelly gas mantles hissed and the coal fires did not draw properly. By contrast, the new science block, in red brick with stone mullions to match the main building, boasted the modern comforts of electric lighting and central heating. In these pristine laboratories Arthur first encountered physics and chemistry.

During Arthur's second year his mother gave birth to a third — presumably unplanned — son, Philip. With more mouths to feed, Sidney tried to get a better headship but without success. They needed a bigger house and moved to a new development on the outskirts of Hessle. Situated in a quiet cul-de-sac, Inglewood, now 14 Marlborough Avenue, is a semi-detached house with a pleasant garden. The household was frugal. They scrimped and economised, clothes were mended, plates scraped clean, yet occasionally Edith managed to employ a uniformed maid. Although she was Yorkshire to the core, she deplored the boys speaking with a broad accent and as an adult Arthur had no trace of Yorkshire vowels. One afternoon, within her earshot, he called to his waiting friends, 'Me Moom says I can coom out' which earned him a reproach.

After sitting the Lower School Certificate (comparable to today's GCSE), Arthur entered the sixth form. Although he could have specialised in the arts, Mr Gore put him on the modern side to concentrate on mathematics and science while continuing with history, English, French and Latin. For Arthur, Latin was the cornerstone of education: 'To walk into a church, ... to read any book intelligently, to catch allusions ... all these need Latin.'[19] Arthur reached his affinity with mathematics after a progression through chemistry and physics. First it was chemical equations that fascinated him, then electricity and magnetism. The enlightened and

---

[19] Letter from Milne to the author, 1 July 1943.

energetic physics master Mr Denham[20] kept his pupils up-to-date with the latest news from the Cavendish Laboratory on atomic structure. 'Electrons and J.J. Thomson were spoken of mysteriously by physics masters with bated breath.'[21] Mr Denham had a marked influence on Arthur, as did Mr Chaffer[22] who taught mechanics, which Arthur rated comparable to Euclidean geometry for its discipline.

Arthur much preferred theory over practical work. Although playing with Atwood's machine or Fletcher's trolley was 'jolly interesting and good fun',[23] he brushed this aside because he felt that the inevitable experimental errors merely cast doubt on the validity of whatever law he was meant to verify. Moreover carrying out experiments added not a whit to his understanding, nor his ability to solve problems. Encouraged by his masters, he digested J.H. Jeans's[24] *Theoretical Mechanics* but struggled with J.J. Thomson's *Discharge of Electricity through Gases* because he had not yet learnt integral calculus. His parents caught him reading the book on a Sunday, thus violating the Sabbath, but he judged it worth their rebuke. At the end of the year he was placed top of his class in chemistry, physics, Latin and French, and in the Oxford and Cambridge Higher School Certificate he gained distinctions in mathematics, mechanics, physics and chemistry.

Hitherto Arthur had shown no particular flair for mathematics, but Mr Gore had a shrewd eye for potential and removed him from lessons in chemistry for more in mathematics, Mr Gore's own subject. He imparted his infectious enjoyment of trigonometry and analytical geometry and

---

[20] Harold Arthur Denham (1878–1921) taught at Hymers from 1905 to 1914. In the war he was made a Companion of the Distinguished Service Order and afterwards became headmaster of Harvey Grammar School, Folkestone.

[21] Milne's tribute to J.J. Thomson in *Nature* (14 September 1940), p. 365.

[22] Clifford C. Chaffer (1884–1939). A Cambridge wrangler, he become head of the Admiralty Research Laboratory, Teddington.

[23] World War II prevented Milne delivering 'The Teaching of Mechanics at School and University'. It was published posthumously in *The Mathematical Gazette* 37 (1954), pp. 5–10.

[24] Sir James Hopwood Jeans (1877–1946). Author of textbooks and popular books. After brief appointments at Princeton and Cambridge, he devoted himself to research, music and writing.

made forays outside the syllabus. Arthur admired how Mr Gore unified scattered theorems 'by an unexpected general method',[25] which probably sowed the seeds of Arthur's lifelong quest to find structured orderliness in natural phenomena.

Arthur's sixth form years were happy, and for his last one he was the head prefect. Besides academic work, he engaged in extra-curricular activities, even finding a niche in sport by reporting on Eton Fives matches and reformulating the rules of the version played at Hymers. He was a prominent member of the Debating Society and adjusted its rules too. It considered issues as diverse as whether there was more entertainment in *Punch* than in Shakespeare, and whether England should build a Channel tunnel. In club procedure he was a stickler for accuracy, probably irritatingly so, and insisted on amending slipshod minutes. He was a keen devotee and secretary of the Scientific Society and gave talks on earthquakes, glaciers and the modern topic of radioactivity.

At the end of his second year Arthur sat the Higher School Certificate again and was placed second in England. By this stage, he had outpaced his father intellectually. Not only did Hymers stretch Arthur's horizons but it dangled the promise of a university education. This was beyond the reach of society's poorer echelons and none of his forebears had been to university. The government did not institute state scholarships until the 1920s, and as A.P. Herbert[26] charmingly put it:

> Nor could the humblest of the island race
> Proceed to Oxford, or the other place.[27]

It was virtually impossible to gain entry to these universities without a scholarship and competition was extremely stiff. The number of Oxford and Cambridge scholarships won by a school was the recognised benchmark of its academic standing and during Mr Gore's first eleven years at

---

[25] Letter from Milne in the *Hull Daily Mail* (19 January 1935).
[26] Sir Alan Patrick Herbert (1890–1971). Writer and humorist. He joined the staff of *Punch* in 1924. Independent Member of Parliament for Oxford University from 1935 to 1950.
[27] Lines from A.P. Herbert's speech in verse on receiving an honorary degree from Oxford University. Herbert, A.P. *A.P.H., His Life and Times*, London: Heinemann, 1970, p. 12.

Hymers, his pupils carried off twenty-seven, thirteen of them in mathematics, which confirmed his and the school's high reputation.

Mr Gore decided Arthur should try for a scholarship at Trinity College, Cambridge. The choice of college was of far greater moment than today because the colleges exercised considerable autonomy.[28] Most lectures were college based, often with a fee attached, and the wealthier colleges provided the best tuition. Trinity was pre-eminent in size and intellectual richness. To be a scholar at Trinity was 'an honour understood in the educational world, in the public schools, in society',[29] and a mathematical scholarship was without compare.

When he set off for Cambridge in December 1913 to sit the exams, Arthur was keyed up to a high pitch of readiness yet filled with trepidation. He travelled with a classmate, Harry Wright,[30] who was also sitting for Trinity and who also had not been outside Yorkshire. On stepping into the two-acre Great Court at Trinity they must have marvelled at its grandeur (Fig. 2). For four days they lived in the college while sitting examination papers every morning and afternoon.

Arthur sailed through the mathematics and science papers, and had extraordinarily good fortune in the general paper. After a compulsory translation of Balzac, there was a choice of questions and two suited him to perfection: 'Heraldry' and 'Parody is one form of criticism'. A third question was 'A hen is only an egg's way of making another egg. Discuss'. Arthur's response went into family folklore for its originality. Somehow he contrived to discuss whether a tiger, an elephant or a square liked sugar best:

> A tiger is a flesh eater and therefore unlikely to like sugar; an elephant is
> herbivorous and might well do so; and a square could not like sugar. But a
> round square, if one exists, must be the exact opposite of a square, and
> therefore might be fond of sugar.[31]

---

[28] King's College provides an extreme example of independence. Until the middle of the nineteenth century King's could present their men for the BA degree without their having to sit Tripos examinations.

[29] Browne, G.F. *The Recollection of a Bishop*, London: Smith, Elder, 1915, p. 178.

[30] Harry Gordon Wright (1895–1983). Industrial chemist.

[31] Letter from G.B. Rimes to the author, 11 February 1975.

A week later the results were out. Arthur was jubilant and one can imagine the rejoicing. He had secured a major scholarship worth £80 a year and his future tutor, Reginald Laurence,[32] told him that he had topped the list with record marks.

Notwithstanding having secured the best scholarship on offer, his university education still hung in the balance. No matter how thrifty he was, £80 was nowhere near enough to cover tuition and lecture fees, books, stationery, food, accommodation, fares and possibly doctor's bills. For a Cambridge undergraduate £160 a year was reckoned to be the absolute minimum. Arthur needed to find more money.

Mr Gore was experienced in dealing with this predicament, and steered Arthur towards likely sources. By examination he picked up a £50 Jameson Scholarship, open to candidates from Hull. The Akroyd Scholarship for boys in Yorkshire schools brought in £50 and the East Riding County Council gave him £60. This made the princely sum of £240, more than the salary of many a schoolmaster. Arthur was also awarded a Hymers Leaving Exhibition worth £50, but since he had already accumulated sufficient funds, it went to Harry Wright, who had secured a £35 award at Trinity:

> I was given the £50 of the Hymers Leaving Exhibition and E.A. Milne's name is on the board. On the strength of my two successes I was given a Hull City Council scholarship of £50, for which Milne living at Hessle was not eligible, so finished with £135, largely due to Arthur's brilliance.[33]

His financial problems solved, Arthur had to clear one last minor academic hurdle. The University of Cambridge, as distinct from its colleges, imposed its own entry requirement, an examination in classics, the 'Little-Go'.[34] He studied Virgil, Xenophon and St Mark's Gospel in Greek, to gain a First in the Little-Go. Finally, all was set fair for going up to Cambridge in the autumn. My grandparents must have been bursting with pride at

---

[32] Reginald Vere Laurence (1876–1934). Historian.
[33] Letter from H.G. Wright to the author, 11 February 1975.
[34] After World War I the university discontinued the Little-Go, but proficiency in Latin at School Certificate remained an entry requirement until after World War II.

their son's lofty academic achievement. Perhaps Edith cocked an imaginary snook at their philistine relations in Hull.

Arthur was ever mindful of his huge debt to the school and to Mr Gore, who watched his career unfold. He was the first Hymerian to be elected a Fellow of Trinity, in 1919, and to the Royal Society, in 1926, occasions on which Mr Gore granted the school a half-day holiday. He gave Arthur a cherished mathematics book passed down from his father. The great Irish mathematician W.R. Hamilton[35] had presented his lecture notes on quaternions to Canon Gore, as he became, and he had had the loose pages bound. In his turn, Arthur bequeathed the book to Trinity College, as a mark of the link between mathematics, his headmaster and the college.

Arthur and Harry took long walks together. Harry's elder brother had been a Cambridge undergraduate, and after he had drowned on the Norfolk Broads their mother wanted Harry to take his place. She advised Edith on the sheets and crockery that Arthur would require and 'everything was sweetness and light'.[36]

In July 1914 Hull was *en fête* to greet King George V who was opening its new docks. Flags flew, bands played, the shops closed, without any hint that the assassination of Archduke Franz Ferdinand would plunge the country into crisis. On Bank Holiday Monday, 4 August, the declaration of war threw Arthur and Harry into a quandary. Fearing that if they joined the army their precious scholarships might be in jeopardy, they consulted Mr Gore. He counselled them to keep to their plans as 'all the air was ringing with rousing assurances'[37] that the war would be over by Christmas.

---

[35] At the age of twenty-two Sir William Rowan Hamilton (1805–1865) was appointed a professor and Astronomer Royal of Ireland. He carved the equations which define quaternions on Brougham Bridge in Dublin.

[36] Letter from H.G. Wright to the author, 11 February 1975.

[37] Montague, C.E. *Disenchantment*, Westport, CT: Greenwood Press, 1922, p. 10.

Chapter 2

The Upheavals of War

The University meets in such circumstances, as it has never known ... We are bound to carry on our work, for by it, we can render definite service to the nation.[1]

Milne was one of some hundred freshmen who arrived at Trinity that autumn. For the first time he was outside the strict jurisdiction of his parents, and he savoured his independence. But his new freedom was sombre. Cambridge was utterly changed by harsh reminders of the war. Soldiers and Belgians blurred the usual demarcation between 'town' and 'gown', and a familiar sight alongside the Union Jack was the Belgian flag of red, yellow and black. After the barbaric atrocities of the Germans who ravaged Belgium and burnt priceless medieval treasures in the library of the University of Louvain, Cambridge offered refuge to its impoverished professors and students.

The city was also full of soldiers billeted in colleges or camped on its outskirts preparing for the trenches. Sad casualties from the battles of Mons and the Marne poured into the hospital set up in the Leys School. Less severe cases were nursed in tents, later replaced by draughty wooden huts, along the Backs, on the site of the present university library.

At Trinity the cloisters of Nevile's Court were made into a makeshift ward with double rows of beds. Whewell's Court was home to the Monmouths, mainly Welsh miners, who brewed their meals out of doors and announced them by trumpet call. Batch after batch of the 5th Officer

---

[1] *The Cambridge Magazine* (10 October 1914), p. 2. The opening words of his address to the university by the Vice-Chancellor, Montague Rhodes James (1862–1936). Medieval scholar. Writer of ghost stories. Provost of King's College, Cambridge, from 1905 to 1918. Vice-Chancellor of Cambridge University from 1913 to 1915.

Cadet Battalion passed through the college and recorded their training in their journal *The Blunderbuss*.

While finding their feet, Milne and Wright kept company. Wanting to contribute to the war effort they applied to the Cambridge University Officers Training Corps, but, to their dismay, were turned down for poor eyesight, 'about which in that war they were very particular'.[2] Instead, they took a first-aid course to qualify as stretcher-bearers for the stream of wounded soldiers arriving in the city.

Usually the boys attended morning service in the College Chapel, where the majestic presence of the octogenarian Master of the College, the Reverend Dr Montague Butler,[3] filled Milne with awe. A relic of the last century, with a flowing white beard, a skullcap and black velvet gloves, he delivered his eloquent sermons in a voice oratorical.

The chaplain, the Reverend J.C.H. How, got the boys interested in the Trinity College Mission, which was attached to St George's Church in Camberwell, a deprived area of London. The mission was founded with the exalted intention to 'civilise the working classes and make them Christians'.[4] Of course the religious and educational programmes could not alleviate the poverty and overcrowded housing, but they instilled an uplifting sense of purpose.

The mission (now Trinity in Camberwell) gave undergraduates the opportunity to lend a hand in the parish, and the chaplain persuaded the boys to take part in 'Men's Service Week' in December after term finished. From his father's work in Hessle, Milne had a fair idea what this entailed. The week brought the boys to London for the first time, and for the last leg of their journey they caught a tram along the Walworth Road. As Milne glimpsed the Doric portico of St George's, perched on the edge of the Grand Surrey Canal, he would surely have identified its Greek revival architecture.

---

[2]Trevelyan, G.M. *An Autobiography and Other Essays*, London: Longmans, Green & Co., 1949, p. 35.

[3]Henry Montague Butler (1833–1918). Classicist. Master of Trinity College from 1886 to 1918. His great nephew, the statesman R.A. Butler, was Master of Trinity from 1965 to1978.

[4]Goldman, L. A. *History of the Trinity College Mission 1885–1985*, Cambridge: Cambridge University Press, 1985, p. 4.

The handsome mission building,[5] designed by no less an architect than Norman Shaw,[6] had cost Trinity the substantial sum of £13,000 and it was a hive of activity. The Reverend H.G.D. Latham,[7] well experienced in mission work, had a team of four clergy and ran a big programme of popular events. The Sunday school classes alone catered for eight hundred youngsters. Given his upbringing, Milne must have had no difficulty assisting with Bible and Sunday school classes. He lodged with a cobbler and visits to men in the parish made him realise that people of little education in menial jobs were fully capable of reading widely and thoughtfully. Later, he served on the Trinity committee which raised money to maintain the mission and pay the clergy stipends.

For his first year at Trinity Milne had rooms in the far corner of New Court on D staircase. Nearby was his tutor, the historian R. Vere Laurence, who dealt with all administrative matters, there being no college office, and who was legally responsible for undergraduates until they attained the age of twenty-one. In spite of documents strewn around his room in apparent confusion, Laurence was competent and trustworthy (so much so, that he was Director of Studies to the Royal Princes, Albert and George, later King George VI and the Duke of Gloucester, when they came up to Trinity). He had a reputation for witty, sometimes malicious conversation. An undergraduate who enquired about his hobbies elicited the prim reply, 'Correcting examination papers'.[8] More to Milne's taste was his moral tutor, the historian Denys Winstanley,[9] an affable, generous host. The shrewd and kindly senior tutor rejoiced in the name W.C. Dampier Whetham.[10] That his dog was called Flush, and his house 'Upwater Lodge' triggered peals of hilarity, Milne not excepted.

---

[5]Today the mission is a modern complex of rooms adjoining a small church.
[6]Richard Norman Shaw (1831–1912). Scots architect, whose buildings include New Scotland Yard, Bryanston School, Dorset, and houses in Bedford Park, London.
[7]After working in London at the Corpus Christi College Mission, Henry Guy Dampier Latham (1868–1944) became Dean of Perth, Australia. In 1911 he returned to England to run the Trinity College Mission.
[8]*Trinity Magazine* (11 June 1914), p. 187.
[9]Denys Arthur Winstanley (1877–1947). Authority on Victorian Cambridge.
[10]Sir William Cecil Dampier (1867–1952). Physicist, agriculturist. When knighted, he shortened his surname. His 1930 *History of Science* (Cambridge University Press) went into four editions.

Milne was taught by world-class mathematicians. His Director of Studies was the Reverend Dr Edward Barnes,[11] an impressive six feet tall, who had been President of the Union. He advised Milne as to the lectures he should attend, including his own on conics, which were models of clarity. He had a simple yet effective technique for handling pupils. Before lecturing, he read out names by category: those whose work was correct, those with a few mistakes, and those who had failed to hand in any answers.

Barnes was one of Trinity's most erudite Fellows. He had entered Trinity as a scholar, was bracketed second wrangler[12] the year Milne was born, and ever since had lived in the college, taking holy orders along the way. A radical theologian, at the liberal edge of the Church of England, he subscribed neither to the Virgin birth nor to the resurrection. He rejected the New Testament miracles because they conflicted with his scientific outlook. When he was a bishop — he never wore a mitre or pectoral cross — he precipitated and survived many an ecclesiastical controversy. Milne admired his steely, uncompromising courage and his endeavours to narrow the gap between science and religion.

Milne was intensely serious about increasing his mathematical knowledge and revelled in his studies. In his second term he sat for a Senior Trinity Scholarship, which could top up his £80 per annum to £100. The examination, set by the college, had three components: an English essay, a general paper and a mathematics paper. Expecting to be faced with problems he had not seen before, he was amazed to spot one that he had already solved. In one of his lectures on algebraic geometry, Herbert Richmond[13] of King's College had set a problem, which Milne explored further by attaching extra conditions. Richmond, a benevolent if somewhat quirky bachelor, took the trouble to sit down with Milne and work the problem through, despite Milne being at another college and not his

---

[11] Ernest William Barnes (1874–1953). Fellow of Trinity from 1898 to 1915. Master of the Temple from 1915 to1919. Canon of Westminster. Bishop of Birmingham from 1924 until his death.
[12] Wranglers, those who gained a First, were placed in order of merit until 1909 when alphabetical listing was introduced.
[13] Herbert William Richmond (1863–1948). University lecturer in mathematics.

pupil. It was this elaborated question, of Milne's own devising, that, to his delight, appeared in the Trinity paper. Not surprisingly, Milne got the Senior Trinity Scholarship, which increased his income to £260.

At the end of the year Milne sat Part I of the Mathematics Tripos.[14] Its broad syllabus included heat, light, elasticity, hydrodynamics and electricity, topics which many other universities assigned to physics. Milne was rewarded for his efforts by a First with distinction. His friends had done well, too: Sam Pollard,[15] whose poor health stunted his career in mathematics, and Harry Thorne,[16] an Australian, whom Milne took home to Hessle for Christmas. Milne was drawn to undemonstrative people, and one of his closer friends was Egon Pearson,[17] shy and reserved, quite the opposite of his formidable father, Professor Karl Pearson,[18] the founder of modern statistics.

Milne's relationship with Barnes, that of a raw undergraduate and a senior Fellow twice his age, mellowed into one of mutual admiration. They shared an interest in mathematical physics and in the compatibility of God the Creator with God the Scientist. By the autumn, however, when Milne returned to Cambridge, Barnes had left for London, to be Master of the Temple. His life at Trinity had become uncomfortable. He had fallen out with his more belligerent colleagues for publicly parading his pacifist principles — in October 1914 he had preached a sermon on the theme 'Love your enemies'— and siding with Bertrand Russell,[19] then a Trinity lecturer, over his disapproval of the war. The Fellows feared that Russell's

---

[14]The Mathematics Tripos is the university's oldest examination.

[15]Samuel Pollard (1894–1945). A gifted draughtsman, he had worked in the drawing office of the Canadian Pacific Railway before coming up to Trinity. After World War I he was a Fellow and lecturer at the College.

[16]Harold Henry Thorne (1890–1953). Born in New South Wales, he gained a degree at the University of Sydney before coming to Cambridge and graduating in 1918.

[17]Egon Sharpe Pearson (1895–1980). Lecturer, Reader and Professor in Statistics at University College London.

[18]Karl Pearson (1857–1936). Statistician, biologist and author of *The Grammar of Science*, Black, London, 3rd edition, 1911. He became a professor at University College London at the age of twenty-seven.

[19]Bertrand Arthur William Russell, Earl Russell (1872–1970). Philosopher and social reformer.

eminence, coupled with his notorious anti-war views, would damage the college's reputation by making it look like a hotbed of pacifism.

With Barnes gone, Milne entered the orbit of G.H. Hardy,[20] the pre-eminent pure mathematician of the first half of the twentieth century, known to countless students through his celebrated *A Course of Pure Mathematics*. Milne was already under his spell, fascinated by his lectures on analysis (the branch of pure mathematics concerned with a rigorous treatment of the theory of functions, especially of the differential and integral calculus).[21] Now he was to be taught by Hardy. Milne counted himself immensely privileged to learn mathematical rigour and style from its supreme practitioner, but with hindsight wished he had had more instruction in technique. Hardy, he felt, carried rigour too far, pursing it as an end in itself, thereby masking the spirit of mathematical discovery. But at the age of nineteen, captivated by Hardy's 'voice of inspiration',[22] Milne set his heart on becoming a pure mathematician and saw endless uncharted vistas stretching before him.

Milne knew Hardy in all his brilliant eccentricity. Hardy regarded God as his personal enemy and harboured vehement likes and dislikes. He hated dogs, mirrors and telephones. He liked cats, crosswords, ball games, factorising taxi numbers, compiling imaginary teams of people in various walks of life, such as cricket, and he encouraged provocative conversation.

Like Barnes and Russell, Hardy openly endorsed an anti- war stance. His great collaborator John Littlewood,[23] also a Fellow of Trinity, was away serving in the Royal Artillery, and his rooms were empty. With his agreement Hardy proposed holding an anti-war meeting in them. The College Council was outraged. It strenuously objected to appropriating college property for such a use and a bitter confrontation erupted. Hardy

---

[20]Godfrey Harold Hardy (1877–1947). Fellow and lecturer at Trinity College. After twelve years at Oxford as Savilian Professor, he returned to Cambridge as Sadleirian Professor.

[21]I am obliged to Leon Mestel for this definition.

[22]Milne, E.A. 'G.H. Hardy' in *Mon. Not. Roy. Ast. Soc.* 108 (1948), p. 44.

[23]John Edensor Littlewood (1885–1977). Lecturer at Trinity College. Commissioned into the Royal Garrison Artillery in 1914. Appointed Cayley Lecturer at Cambridge in 1920 and in 1928 Rouse Ball Professor.

drummed up support for a protest against the council, but nothing was resolved and dissension split the Fellows.

Milne was contemptuous of lecturers who did not measure up to his high expectations and even Hardy fell from grace. In Hull during the long vacation, Milne had systematically worked through a mathematics book he found in a library. He was therefore greatly disappointed the following term to hear Hardy quoting from it, word for word. Milne approved of the clear explanations of H.F. Baker,[24] the leading British geometer, but deplored his indistinct tiny writing on the blackboard and his rapid delivery. Since he never paused for breath it was impossible to take everything down. As Milne put it, 'If you sneeze twice you lose a page of notes'.[25]

In January he attended lectures on hydrostatics by Sydney Chapman,[26] whom he judged rather nervous. Quite the opposite of Baker, Chapman considered his audience. He waited for his listeners to keep up with him and his delivery was good. 'I had a small discussion with him this morning in the lecture. I am told that this is rather unusual, as he won't let you argue with him.'[27] But during Milne's supervisions with Chapman, who was barely eight years older, Chapman eased up, and their encounters led to an enduring friendship. Chapman had a background in engineering, having been to technical school in Manchester and gained a degree there before taking up mathematics in Cambridge.[28] He exerted a profound influence on Milne's perception of the purpose of mathematics with an attitude that was completely at odds to Hardy's. Whereas Hardy insisted that mathematics was an elegant exercise, not to be contaminated by application to practical situations, Chapman regarded mathematics as a tool for solving problems.

---

[24] Henry Frederick Baker (1866–1956). Professor of Astronomy and Geometry 1914 to 1936.
[25] Letter from Milne to G. Milne, 20 January 1916, in the author's possession.
[26] Sydney Chapman (1888–1970). Geophysicist. He held chairs at Manchester, Imperial College London and Oxford. An enthusiastic cyclist, he covered enormous distances, sometimes formally dressed for an evening function.
[27] Letter from Milne to G. Milne, 20 January 1916, in the author's possession.
[28] The progression from Manchester to Cambridge was a well-trodden route, travelled by the applied mathematician Sir Horace Lamb, the physicist Sir J.J. Thomson, the astrophysicist Sir Arthur Eddington and others.

Chapman was a Quaker and, on religious grounds,[29] a conscientious objector. So it happened that all three, Barnes, Chapman and Hardy, to whom Milne was answerable for his studies, defied the prevailing patriotism. This contrasted with Milne's earlier mentors, his headmaster and his father, who supported the war, as did his family. Milne's favourite cousin Clifford left his job at a bank to enlist, and Geoffrey, after winning a scholarship to read chemistry at Leeds University, joined the King's Own Yorkshire Light Infantry. It cannot have escaped Milne's notice that his Trinity teachers, who made no secret of their convictions, suffered considerable social discomfort living within a closely-knit yet hostile community. Their plight must have registered in the recesses of Milne's mind, perhaps against the day of his own controversies. Notwithstanding these unforeseen incongruities, Milne was entranced by the serenity of Trinity, and dazzled by its splendour. He always spoke of it with the greatest fondness and it would ever be his spiritual home. At this stage, in the middle of his second year, he left the college at short notice because the war unexpectedly immersed him in a mathematical education of quite a different character, with important consequences.

---

[29] By World War II Chapman had modified his idealism and worked for the Army Council.

Chapter 3
Adventures with Reflections

There were already a few anti-aircraft guns, but no proper scientific studies had been made of their use, and, as it turned out, most of what was supposed to be known was wrong.[1]

At the outset of World War I the British Army distrusted civilian scientists, sceptical of their making a useful contribution to victory, whereas in Germany science, scientists and technology commanded respect. English military personnel perceived science as the preserve of slightly eccentric, wealthy gentlemen tinkering away in private laboratories, so attempts to improve artillery by scientific means were dismissed as unwanted and unwarranted intrusions into military expertise. Given this complacency, it is not surprising that research for warfare evolved in an *ad hoc* manner, without the systematic allocation of scientific expertise that there would be in World War II. Projects sprang from individual initiative, and depended on personal networks of contacts for recruiting suitable teams. Having the ear of people in high places might be crucial to gain government support.

The chain of events that swept Milne into the research that would revolutionise gunnery had its origins in Trinity College, and Milne was fortunate to be in the right place at the right moment. His extraordinary adventures over the next three years were the springboard to his accelerated life in mathematics.

The story begins with the inventor-engineer Horace Darwin[2] (Fig. 3), the fifth son of Charles Darwin. A former Fellow of Trinity and a Fellow

---

[1]Hill, A.V. *Memories and Reflections*, p. 124, AVHL I.
[2]Sir Horace Darwin (1851–1928). Civil engineer. Mayor of Cambridge in 1896. My father would be astonished that he and Darwin share great-grandchildren through the marriage of my daughter Miranda to Horace Barlow.

of the Royal Society, he had a gift for innovation and had built up a flourishing business designing and manufacturing scientific instruments and equipment. His Cambridge Scientific Instrument Company supplied the armed services with hundreds of gauges and artefacts such as dynamometers for testing aeroplane propellers.

Darwin was keenly interested in aeronautics and realised that knowing the height of an aeroplane from the ground was vital if a shell were to be launched against it. That the height needed to be measured was a notion that military die-hards greeted with cynical disbelief. They failed to foresee the impact of aviation on warfare. Until World War I, battles were confined to land and sea, and none took place in the air. The third dimension was a new combat zone. In 1914, only five years after Louis Blériot crossed the Channel in his monoplane, flying was still a flamboyant, glamorous sport of reckless races and daredevil stunts. At the beginning of the war planes only undertook reconnaissance missions, but as soon as they went on the offensive and dropped bombs, the need arose to shoot them down. A new type of gunnery, anti-aircraft gunnery, was born and Milne was one of its pioneers.

In early 1916 Darwin wanted to assess the usefulness of a height-finding apparatus he had in mind and consulted a recent Fellow of Trinity, Archibald Hill[3] (Fig. 4), known as AV. The job of refining the height-finder demanded initiative, a knack for experimentation and a sophisticated level of mathematics, qualities possessed in abundance by Hill, whose outstanding work in muscle dynamics would earn him a Nobel Prize.[4] Athletic, handsome, with prematurely greying hair, Hill had an aptitude for engineering beyond his exceptional prowess in mathematics and science. Clearly his wartime duties as a brigade musketry officer were not commensurate with his profusion of talents.

---

[3] Archibald Vivian Hill (1886–1977). Brackenbury Professor of Physiology at Manchester from 1920 to 1923. Professor at University College London from 1923 to 1951. Independent Member of Parliament for Cambridge University 1940 to 1945. Nobel Laureate 1922.

[4] The following year, Hill's marriage to Margaret Keynes took him into the fringes of the vast Darwin dynasty. One of her brothers was the economist John Maynard; the other, Geoffrey, a surgeon, married one of Horace Darwin's nieces.

The upshot of their collaboration was a device, known as the Darwin–Hill Mirror Position Finder, which depended on triangulation. In simple terms, the position finder consisted of two horizontal mirrors, ruled with a grid of coloured lines like graph paper, placed a good mile apart. Observers at the mirrors caught the reflection of an object, such as a plane, high up in the sky and both noted the position of the image on the grid. They communicated by field telephone, and a trigonometric calculation, which involved the x/y co-ordinates of the image, together with other information, produced the height of the aerial object. In actuality the procedure was considerably more complicated.[5]

Hill's first step in refining the position finder was to recruit a team, and he approached G.H. Hardy. Hardy, though, had no intention of compromising his opposition to the war by applying his brains to research, and recommended another Fellow of Trinity, the mathematician Ralph Fowler,[6] and Milne, 'one of his best pupils'.[7] Glad to be able to contribute towards the war effort, Milne immediately abandoned his studies and plunged willingly into the enterprise. He must have wondered where this unforeseen turn of events would lead and cannot have had the remotest idea of its far-reaching influence on his life.

Fowler, currently an artillery officer, had received a shoulder wound at Gallipoli and agreed to join Hill during his convalescence in Cambridge. Large in stature and an accomplished golfer and tennis player, Fowler had a first-rate mind, an engaging, forceful personality and a loud voice. A committee was said to be 'strong' if he were on it. Hill, aged thirty, and Fowler, twenty-seven, were men of authority, and Milne, slight of build, a mere undergraduate just turned twenty, felt much their junior. He energetically applied himself, whether wrestling with mathematics or scurrying about performing menial tasks such as driving posts into the ground. He noticed how Fowler's abstract mind complemented Hill's experimental

[5] See Van der Kloot, W. 'Mirrors and Smoke' in *Not. and Rec. of the Roy. Soc. of London* 65, 4 (20 December 2011), pp. 393–410.

[6] Sir Ralph Howard Fowler (1889–1944). Mathematical physicist. Plummer Professor of Mathematical Physics at Cambridge from 1932 to 1944. In World War II he co-ordinated British and North American research programmes. He married Rutherford's daughter, Eileen.

[7] Letter from A.V. Hill to W.H. McCrea, 2 November 1950, EAM, MS.Eng.misc.b.429.

insight to create a most effective partnership. Milne marvelled at how they set about tackling a problem, made inferences, exerted judgements and came to conclusions. They could not have given him a better training in how to carry out research.

The trio, the nucleus of a team that ultimately grew to a hundred people, made preliminary tests with the height-finding device by estimating the heights of telegraph poles along the Huntingdon Road. The mirrors, two feet square and a quarter of an inch thick, were made by Darwin's company. They were supported by tripods and the distance between them, the baseline, was carefully measured. The winter was exceptionally severe so that making painstaking adjustments with numb fingers to level the mirrors, and marking the images of the poles in special ink on paper laid over the mirrors, demanded fortitude and patience. But the results were most gratifying, accurate to within one percent.

The Darwin–Hill Mirror Position Finder had distinct advantages over other height-finders. Firstly, its design incorporated a self-checking system, based on the y co-ordinate, which ensured simultaneity of observation, essential for a moving object. Secondly, it gave a larger field of view than scanning with a theodolite; and thirdly, the observer was more comfortable because he did not have to crane his neck upwards.

Hill's next step was to test the device on a moving, rather than station-ary, target. He negotiated with Northolt Aerodrome, on the western out-skirts of London, for them to run flights for testing, and took his team there in March. (Hill and Fowler slept in wooden sheds supplied by Darwin's company, while Milne found digs in Ruislip.) They draped quantities of telephone wire around the perimeter hedges and set up mir-ror stations at either end of the airfield to give a longer baseline for increased accuracy. Things did not go well, however, because the Aerodrome was not co-operative and did not run many of the promised flights. Nevertheless, they struggled on and Hill demonstrated the height-finder to officials from the Munitions Inventions Department (MID).[8] Yet, 'the attitude of the authorities was chilly, there was apathy from the

---

[8]The remit of the Munitions Inventions Department ranged from artificial limbs to patents, camouflage and chemical waste. It vetted 48,000 new ideas, most of them impracticable.

people at the Aerodrome, who regarded us as a set of rather uninteresting cranks, wanting to make a science of a thing which was intended by nature to remain a sport'.[9]

At this point the project seemed likely to founder. Hill searched for a less discouraging place to work but nothing presented itself. He ran into financial difficulties with the MID, which raised objections to paying Milne as it expected him to subsist on his scholarship money although he was no longer studying. Hill protested and got Milne the handsome wage of 25 shillings a week. Then Hill pleaded his own case as he had not been paid for two months, nor reimbursed expenses of a few pounds for graph paper, a large protractor and a T-square.

Worse still was the threat of losing Fowler. The Admiralty objected to his joining Hill permanently and wanted him to be an inspector of steel — about which he knew almost nothing. At this critical juncture, when the project was on the brink of collapse, Horace Darwin intervened. He knew Arthur Balfour,[10] the First Lord of the Admiralty, whose younger brother Francis[11] had been his Trinity contemporary and close friend. Arthur Balfour was well disposed towards scientific research and a word with him ensured that Fowler remained.

For the moment Horace Darwin's personal intervention saved the enterprise, but there was no clear slot for Hill's outfit in the diffuse framework of national defence. Responsibility was split between the army, the navy and rival government departments. The War Office controlled weapon design while the Ministry of Munitions had charge of the proving grounds at Woolwich. Home defence was shared between the Royal Naval Air Service, the Royal Flying Corps and the Home Defence Corps, but had no single command head. (The RNAS and RFC merged to form the RAF in April 1918.) Aerial defence was shuttled between the War Office and the Admiralty. The Board of Invention and Research, nicknamed the 'Board of Intrigue and Revenge', assessed anti-aircraft weapons but had no defined relationship with the Admiralty. Moreover, the Admiralty did not

---

[9]Hill's report for events up to 28 September 1916, p. 12, AVHL I 1/37.
[10]Arthur James Balfour (1848–1930), first Earl of Balfour. Prime Minister from 1902 to 1905.
[11]Francis Maitland Balfour (1851–1882). Naturalist. He died on Mont Blanc soon after Cambridge University had created a chair of animal morphology for him.

permit the board to communicate with the army and the Royal Flying Corps. Such was the confusing tangle of bureaucracy that confronted Hill.

In addition, he was up against military prejudice to a scientific approach towards warfare. He had to grapple with widespread ignorance about high-angle fire because army personnel were reluctant to accept that it differed from field gunnery. When a shell passes through the upper atmosphere it encounters changes in temperature, pressure and winds, and these must be taken into account to predict its path. The policy of just giving extra elevation to a gun was fundamentally flawed because 'height' is a more important parameter than 'range' in high-angle gunnery. Since they were oblivious to this crucial distinction between field gunnery and anti-aircraft gunnery they were indifferent to the value of the height-finder and to the necessity of designing guns specifically for high elevation. Add to this that fuses were unreliable, sometimes killing the gun crew, and it is apparent that early anti-aircraft gunnery was literally a hit-or-miss affair.

Hill thought he might be able to continue at Shoeburyness, on the Essex coast, where Major Douglas H. Gill took an unusually enlightened interest in gunnery. But Brigadier General Walter D. Sclater-Booth objected because he thought that instruction and experimentation did not mix.

In May, to Hill's relief, the physicist Richard Glazebrook[12] came to the rescue. Chairman of the Advisory Committee of Aeronautics,[13] he was influential in government circles and his support signified conspicuous approval. He was Director of the National Physical Laboratory (NPL) at Bushy Park, Teddington, which tests instruments and materials and lays down criteria for standardising physical units. The NPL already carried out research in aeronautics and Glazebrook generously put its facilities at Hill's disposal, along with a lecture room for the team's exclusive use. Delighted by this invitation, Hill transferred operations to

---

[12]Sir Richard Tetley Glazebrook (1854–1935). First Director of the National Physical Laboratory in Bushy Park, the former home of Dora Jordan and the Duke of Clarence (later William IV).

[13]Other members of the committee were Horace Darwin; Mervyn O'Gorman, superintendent of the Royal Aircraft Factory, Farnborough; F.W. Lanchester, designer of the Lanchester car; and W. Napier Shaw, Director of the Meteorological Office.

Teddington, bringing his team with him, which had now grown to five people.

Milne had digs close to Bushy Park, at 57 Coleshill Road, which he shared with another undergraduate, Arthur Hawkins,[14] recently recruited by Hill. Milne was a voracious talker and therefore disappointed to find Hawkins taciturn, but at least Hawkins was a companion of his own age, and in their free time they explored the area by bicycle, zigzagging back and forth over Thames bridges. Milne was an energetic cyclist and made a round trip of twenty-five miles to Ruislip to retrieve his laundry from his landlady.

Milne was totally committed to the project, and excited by it. Trials with the height-finder spawned mountains of tedious number crunching and he romped through the calculations with a slide ruler and tables. Before long he acquired a hand-cranked Brunswiga machine, known as a 'crasher', which he treated warily, and claimed to feed with paper clips. Excursions to London with Hill and Fowler for meetings at Spring Gardens, next to Admiralty Arch, made a welcome diversion from 'the fog of sines, cotangents, logarithms, azimuths, altitudes, angles of sight, subsidiary bases, least squares, and plotting diagrams'.[15] In a cocky mood Milne wrote:

> Another thing ... requires us to look up the cotangents of all angles from 20° to 60°, add them together in pairs, in every possible way, look up the logarithm of the sum and then multiply it by 25. I went nearly dotty yesterday ... I got to be able to subtract 2 numbers of 6 figures each and call out the result quicker than the other people could write it down.[16]

In the NPL's happy environment the work gathered momentum. Edwin Rayner,[17] a Cambridge-trained engineer in charge of electrical gearing at

---

[14] Arthur Charles Hawkins (1896–1972). After the war he completed his degree and taught at the Perse School, Cambridge.

[15] Letter from Milne to G. Milne, 26 May 1916, in the author's possession.

[16] Letter from Milne to G. Milne, 20 May 1916, EAM, MS.Eng.misc.b.423.

[17] Edwin Hartree Rayner (1875–1963) became superintendent of the NPL's Electricity Division. His sister Eva was William Hartree's wife.

the NPL, was most hospitable and invited Milne home for family meals. The Rayner children treated him as a young uncle, and after the war the Rayners included him in their seaside holidays. At Rayner's suggestion, Hill invited Rayner's brother-in-law and cousin, William Hartree,[18] formerly a Cambridge lecturer in engineering, to join the team. He proved a great asset, always cheerful and ready to turn his hand to whatever was needed. He devised a simpler form of height-finder, which was cheap to produce and easy to set up. The calculations avoided trigonometry, which made it suitable for men at the front, and the following year twenty Hartree Height Finders went to France.

The Mirror Position Finder was now put to routine use to evaluate other height-finding devices, some with extraordinary names like 'Flower Basket' and 'White Elephant'. For a crude initial assessment Milne would clamber up to the roof of the building and wave a white card as a sighting target.

The team made excellent progress. Hill was ready to show off the device to a formidable array of senior personnel in trials that involved balloons and kites at the NPL and at Hyde Park, for which he obtained police permission. The climax would be at dusk when the team would ascertain the height of a plane lit by searchlight. As the day of the demonstration approached, the pace quickened, and Milne was at full stretch working late.

On Wednesday evening Hill came in from London with the news that half the world was coming to see all we could show them ... The Director General of Munitions Designs, with his brass-hatted staff, the Comptroller of the MID, Horace Darwin, Sir Alexander Kennedy[19] ... [are] to arrive for a demonstration in daylight from 6 to 8 on balloons and an aeroplane, with a similar affair to come off at 10:30 in the dark.

---

[18]William Rayner Hartree (1870–1943). Engineer. After the war he learnt to dissect tortoises, frogs and hedgehogs to assist Hill in his research on the dynamics of muscles.

[19]Sir Alexander Blackie William Kennedy (1847–1928). Professor of Engineering at University College London from 1874 to 1889, where he introduced the method of teaching students by means of practical experiments. He designed generating stations at Edinburgh, Glasgow and Manchester.

Most complicated arrangements had to be made for the night work. Various slits in mirrors and sighting wires have to be carefully illuminated with concealed electric lights ... provided with batteries and switches and if ... we had to run two or three lamps off one set of accumulators in parallel, then each lamp has to have its own adjustable resistance ... In the morning I had to cycle over to Twickenham to buy up all the electric torches I could lay hands on. ... The maze of wires ... was appalling. ...

The heights calculated from the readings of different instruments agreed absolutely perfectly. The airman, the Commander of the Hounslow Aerodrome ... finished up by doing marvellous banks and turns just above the heads of the crowd, finally standing up and raising his hat to them.[20]

The trials were a huge success and Milne was elated that Hill cited his work and recommended him for a pay rise. The Darwin–Hill Mirror Position Finder hovered on the threshold of official recognition and the next demonstration removed any lingering doubts. Hill persuaded the Ordnance Board to let him track the trajectories of high-angle shells fired by 6-inch AA guns across the estuary at Great Yarmouth. The chief purpose was to evaluate a rapid new method of calculating high-angle trajectories, devised by Hardy's colleague, Lieutenant John Littlewood. Until then, a time-consuming step-by-step 'method of small arcs' had been applied, and, if satisfactory, Littlewood's method would be much more efficient.

To track a shell required an extra man. Linked by field telephone to the mirror observers, he stood at the gun and recorded the time that elapsed between the moment the gun fired the shell and the moment it burst.

Hill's team, now grown to eight, established mirror stations at Great Yarmouth and Gorleston-on-Sea. Unfortunately the telephone wires they strung along the lamp posts often gave way, and while William Hartree was making repairs he was accosted at bayonet point on suspicion of 'signalling to the enemy'. Bad weather caused frustrating delays, exacerbated by the first fine day being a Sunday, a day on which the army refused to allow firing. When the trials finally took place the results were devastating for the army, as they plainly showed that their firing tables were wrong.

---

[20]Letter from Milne to G. Milne, 28 July 1916, in the author's possession.

For everybody else the outcome was highly favourable: 'To the astonishment and joy of all concerned the observed positions of the shell bursts
fell exactly on Littlewood's trajectory.'[21] In the widely circulated report,
Milne was the proud author of the appendix. Cock-a-hoop, he dashed off
a parody of the report for his brother's amusement, awash with acronyms
and technical jargon.

The trials were conclusive and the Mirror Position Finder was finally
accepted. All that Hill needed now was a permanent base. If the army
was not well disposed towards him might the Royal Navy be more
accommodating? A promising overture from Captain Vincent L. Bowring
of the Royal Navy Gunnery School at HMS *Excellent*[22] on Whale Island,
Portsmouth, prompted Hill and Fowler to take the equipment there. They
departed full of hope, leaving Milne with a pile of numerical work. 'For
three days we worked like slaves at the calculations,'[23] and on their return
Milne sensed an air of expectant excitement.

> The [naval] officers were extremely keen on the experiments. ... Altogether
> 150 rounds were fired, but the remarkable thing was that over 50 of them
> were 'blind', i.e. the fuses did not go off. About the remaining 100 bursts,
> every blessed thing is known: actual position, time of flight, time taken by
> sound to reach observers, deviation from angle of sight, wind blowing at the
> moment in the neighbourhood of the burst, elevation of gun, fuse-length,
> barometer, thermometer, and finally the muzzle velocity of the gun.[24]

Their suspense was short-lived and Captain Bowring cordially invited Hill
to install his team at HMS *Excellent*. After six months of uncertainty and
setbacks, to obtain such an ideal workplace — the country's top gunnery
school — was a triumph and a tribute to Hill's perseverance and charis-
matic leadership.

Buttressed by government endorsement Hill was able to expand his
team. The curious but remarkably harmonious mix of seasoned dons,

---

[21]Milne's obituary of R.H. Fowler in *Obit. Not. R. Soc.* 5 (1945), p. 66.
[22]In 1985 HMS *Excellent* was closed and absorbed into HMS *Nelson*.
[23]Letter from Milne to G. Milne, 13 August 1916, in the author's possession.
[24]Letter from Milne to G. Milne, 18 August 1916, in the author's possession.

some of the country's best mathematicians, and young inexperienced undergraduates, was designated the Anti-Aircraft Experimental Section of the Munitions Inventions Department. Unofficially they were known as Hill's Brigands.

By this stage, in late August 1916, the war was going badly. The unremitting battle of the Somme was in deadlock, and prowling U-boats caused brutal shipping losses. Portsmouth, home to many naval families, was in mourning for the thousands of sailors who perished in the disastrous battle of Jutland. As for aerial warfare, our nippy manoeuvrable planes like the Bristol Scout disposed of the ponderously slow, giant Zeppelins, but soon the Germans dispatched planes on bombing raids. There was an urgent imperative to make our AA guns more effective.

While most of the team transferred to Portsmouth, Milne remained briefly in Teddington immersed in the trajectories of shells. (He had already notched up a resounding success by resolving a puzzling anomaly about a correction applied to fuses.) Milne was working alongside his former lecturer Herbert Richmond, one of Hill's latest recruits. They were engrossed in the mathematics of wind and struggling to make a correction for high-angle shells that involved summarising winds blowing at different heights with different speeds in different directions.

Milne liked the challenge of exploring a relatively uncharted subject and the problem intrigued him:

> I have eaten, drunk, and slept through winds. First, to my great surprise I worked out a general formula for wind effect on a shell fired vertically. ...
> The integrals one arrives at and the new functions one has to tabulate are really magnificent.[25]

The breakthrough came with their invention of the 'equivalent constant wind'. By an elegant simplification, they replaced the complicated array of winds with a single fictional wind that had the identical displacement on the shell. The concept proved extremely valuable, went into the literature and continued to be applied in World War II.

---

[25]Letter from Milne to G. Milne, undated, EAM, MS.Eng.misc.b.423.

Captain Bowring welcomed the Brigands to HMS *Excellent* and ensured that guns were fired according to their wishes. Milne admired his forthright language — he uttered bloodcurdling oaths about what he would do to any German airman landing on Whale Island. He allocated the Brigands a large attic with a big table covered in linoleum, hitherto used for polishing the Mess silver, suitable for plotting trajectories. They designated it the Ballistics Office and it was their nerve centre. Here they meticulously planned trials and analysed results. There was enough space to organise their operations and Milne, a stickler for orderliness, was impatient with anybody who was careless: 'Loud are one's vituperations when one comes across a paper without the name of the gun.'[26]

Permanent mirror stations connected by post office telephone lines were set up at Eastney and four miles away at Hayling Island. The Brigands systematically investigated the host of factors which determine path, motion, spin and yaw of a shell.[27] A year later their revised firing tables were applied to 'every anti-aircraft gun in either service'.[28] The height finder had a valuable application to meteorology. By observing the drifts of smoke puffs from a shell the Brigands deduced information about winds, vital to pilots. To quote the official history of the Ministry of Munitions, 'It would hardly be an exaggeration to maintain that the whole science and practice of artillery has been revolutionised by the development of AA gunnery.'[29] The Brigands won international renown for their scientific approach to warfare and their methods were copied by other countries.[30]

The work, which Hill directed from London, spread to numerous locations: NPL for the dynamics of shell flight; University College London for computations and experiments on fuses; Rochford and Stokes Bay for sound locators. The Travelling Ballistics Party effected the calibration of guns in England and France.

---

[26]Letter from Milne to G. Milne, 13 December 1916, in the author's possession.

[27]For a technical evaluation of the Brigands' work on high-angle fire see David, T.R.V. *British Scientists and Soldiers in the First World War*, PhD thesis, London: Imperial College, 2009, pp. 126–196.

[28]MUN 5/117/700/4 Appendix A, p. 2, NA.

[29]*History of the Ministry of Munitions* 10, Part VI, London: Ministry of Munitions, 1914–1922, p. 23.

[30]Skentlebery, N. *Arrows to Atom Bombs*, London: HMSO, 1975, p. 193.

Fowler took charge at Whale Island. On occasion his over-zealous bustling manner created friction, and Milne deplored Fowler's disregard for keys and other people's fountain pens, including his own — Richmond bought him a new one when his original pen was lost. Yet Milne could not help admiring Fowler's dynamic personality, which he portrayed in a breezy sketch:

> He bullied the C's-in-C., browbeat the Admiralty, ... Shoeburyness trembled and cursed, Woolwich pleaded, Portsmouth stood aghast. There was hardly a gun or fuse that didn't eat out of his hand.[31]

The Brigands saved the lives of countless gun crews by unravelling the mysteriously erratic behaviour of fuses. Experiments revealed that the predominant factor determining the rate at which a fuse burns (which controls the timing of the explosion) is the angular velocity of the shell at the instant it leaves the gun. This, in turn, depends on the rifling, the spiral groove within the gun's barrel. To compare three sizes of rifling was a hazardous business, as Milne described:

> We had an exciting time shooting yesterday. Shells set to burst at 20 or 30 seconds burst after 5 seconds, and two were premature, i.e. burst 50 yards or so off the muzzle. We had to send a seaman out to search for bodies, dead or alive. No one, however, was bagged.[32]

Reducing the rifling made fuses reliable, 'blinds' a rarity, and gun crews safer.

Another worry was that guns wore out: the higher the muzzle velocity the faster the erosion. Guns needed re-lining after firing 1,200 to 1,500 rounds and replacing after 3,000 rounds. The Brigands specified new designs for AA guns, which not only reduced the muzzle velocity, but rendered the guns more powerful. By 1918 a larger shell of 16 lbs (7.3 kg), compared with 12.5 lbs (5.7 kg) previously used, had an increased range and could reach a height of five miles.

---

[31]Milne's unsigned article 'Those in Authority' is a sketch of Fowler as Senior Proctor, *Granta* (25 May 1923), p. 469.
[32]Letter from Milne to G. Milne, 6 December 1917, EAM, MS.Eng.misc.b.423.

Milne was a pivotal member of the group and ballistics matured him as a mathematician. Apart from number crunching and gun trials, a variety of other jobs enriched his experience. He attended meetings in Westminster with the Ministry of Munitions, consulted Littlewood at Woolwich Arsenal, went to the NPL and took instruments for calibration to Darwin's workshops in Cambridge. Taken together this gave Milne a comprehensive idea of the scope and impact of the Brigands' activities — unlike the strict secrecy established in World War II when each unit was a small cog in the whole — and led Milne to form a low opinion of army officers:

> The English officer is neither a mathematician nor a scientist ... Bit by bit we are taking the control ... away from the officer, and putting it on instruments which work automatically. The responsibility ... on the officer is to keep his men thoroughly trained and to keep his instruments in adjustment. The latter he will wriggle out of if he possibly can. ...
>
> We suggest the principles of designing new guns, new fuses, new methods of fire, and point out the faults in the present ones. But we are a full 9 months ahead of what is carried into practice. The height-finders we played about with at Teddington last summer have this summer been installed all over England.[33]

That Milne was thrust into the company of senior mathematicians and learnt from them at close hand was his greatest good fortune. Besides Richmond, he had contact with some of Cambridge's most renowned lecturers: G.T. Bennett,[34] E.G. Gallop[35] and R.A. Herman.[36] It was not such

---

[33]Letter from Milne to G. Milne, 30 October 1917, EAM, MS.Eng.misc.b.423.

[34]Geoffrey Thomas Bennett (1868–1943), lecturer at Emmanuel College, held speed records for bicycling, lying flat and using a mirror to see ahead, to bring the racing results from Newmarket to Cambridge.

[35]Edward Gurney Gallop (1862–1936). Lecturer and Fellow of Caius College. He bequeathed land known as 'Gallop's Pieces' to Wicken Fen, Cambridgeshire, for a nature reserve. Joint author of a celebrated paper on spinning shells, *Phil Trans. A* 221 (1921), pp. 295–387, with Fowler, Richmond and Christopher Noel Hunter Lock (1894–1949), scholar of Caius. After the war, Lock joined the National Physical Laboratory.

[36]Robert Alfred Herman (1861–1927). A brilliant lecturer and coach, he was one of Milne's examiners for Part I of the Tripos.

good luck that he was landed with the unenviable task of instructing the formidable Professor Karl Pearson (known as KP), aged sixty, who ran the Galton Laboratory of Biological Statistics at University College London. Hill had wanted to engage Pearson's son Egon to cope with a mass of calculations from the gunnery trials but he was not available and his father offered for himself and his department to take on the huge task of drawing up new trajectory charts and tables for every AA gun in the army and navy. The intricate calculations took account of range, elevation, muzzle velocity, air temperature, wind speed, weight, shape of shell and much else. Given Pearson's famously pugnacious and domineering character, Hill knew negotiations would be delicate, and in choosing Milne he was well aware that he was putting Pearson in the invidious position of receiving instructions from someone young and inexperienced. Tactfully, Hill introduced Milne as a 'nice little modest person, very clever and rather shy'.[37]

Everything began civilly enough on 2 January 1917. They worked late into the evening and Pearson took Milne home for the night to his house in Hampstead, at 7 Well Road, which now bears a blue plaque commemorating Pearson. For the next few days, while initiating Pearson into the complexities of the tables, Milne was his houseguest, then he moved to a college hall. Once back in Portsmouth Milne plied Pearson with pages of figures.

Hill pressed Pearson to produce the tables quickly and Pearson shelved all his own work. By the autumn he was huffy. He simmered with objections to the calculations expected of him and his laboratory, and feared for its reputation. He complained that imperfect data and inaccurate formulae led to annoying discrepancies and he baulked at methods which he regarded as rough and ready. Milne and Richmond failed to quench his anger and Pearson threatened to sever all connection with the Brigands. Even Hill could not pacify Pearson and, at crisis point, it took the controller of the MID, Colonel H.E.F. Goold-Adams, to persuade Pearson to finish the monumental task. Throughout this exacting period Pearson made not a single entry in his laboratory journal.

---

[37]Letter from A.V. Hill to K. Pearson, 31 December 1916. Pearson papers, University College London Library, Special Collections.

A pleasanter task was a memorable excursion to Benson, near Wallingford, to consult William Dines,[38] the leading authority on the upper air. He was the inventor of the pressure tube anemometer that measures both the speed and direction of wind. One fine day in May 1917 Milne and Fowler set off from Whale Island on their bicycles and at Portsmouth station stowed them in the guard's van of a train to Reading, whence they pedalled over the Chilterns to Benson. 'The country was topping, heaps of bluebells in the woods, lilac, chestnut and apple blossom everywhere.' Dines, tall, spare, austere, was most helpful and promised to lend them equipment. At the end of a long day, returning as they went, having covered some thirty miles in the saddle, they 'came home rejoicing',[39] cycling in the blackout without lamps from Portsmouth station back to Whale Island.

Milne had a warm affection for the island. As soon as he crossed the swing bridge past the sentry he felt absorbed into the *Excellent*'s vibrant community. In his snug little 'cabin', with a coal fire tended by a seaman, he had a bunk bed with drawers below and a washbasin that folded into the wall. Civilian Brigands were treated without discrimination exactly the same as Brigands in uniform. In the Mess they ate together with the *Excellent*'s officers under the watchful gaze of naval portraits on the walls. One of these was the familiar face of the sailor on a packet of Players cigarettes, whose cap ribbon was changed from HMS *Excellent* to HMS *Hero*. On guest nights, the Mess silver gleamed resplendent on the tables, a band played in the gallery, and the dinner concluded with a 'sing-song' or wild games among the billiard tables.

When Milne's friends were in the vicinity, he took proprietorial pride in showing them around Whale Island. Remarkably, it was artificial, made twenty-five years earlier by convicts who dumped load after load of mud excavated to create Portsmouth harbour. Since the mud shifted with the tides, the structures on Whale Island were only two storeys high. Besides naval buildings, the island had a cricket pitch, a rose garden, a duck pond and aviary, a farmyard of chickens and pigs, and even a small zoo of wallabies and monkeys brought home from foreign shores by sailors.

---

[38]William Henry Dines (1855–1927). President of the Meteorological Society from 1901 to 1902. His father George Dines invented the hygrometer.

[39]Letter from Milne to G. Milne, 24 May 1917, EAM, MS.Eng.misc.b.423.

On Sunday evenings Milne withdrew to the *Excellent*'s library to indulge his love of letter writing. He recounted firing sessions that went on until 8 pm, which in November was cold work, 'standing in half a gale of wind, without a hat, with a telephone head piece on'.[40] He was cautious about what he told his parents, but, in sprightly banter laced with puns, hid nothing from Geoffrey and instructed him to keep 'squat', that is secret, the technical information he divulged. It was probably in the library that Milne composed 'The Brigands' Lament', a rollicking satire on the tribulations of gun trials which parodied Edward Fitzgerald's translation of *Rubaiyat of Omar Khayyam*, then much in vogue. These quatrains give the flavour:

> Musing o'er Figures, Tangled and Awry,
> I hear a Voice from near the Deck to Cry,
> 'Get up, my People, up and Carry on,
> Before this Weather has had time to Fly.'

> But lingering not I speed me through the Mire
> To Eastney, where I view (in speechless Ire)
> Depending from the Gantry by its Chains,
> The Gun the Party hastened out to Fire!

> The Mirror Parties try to make a Trace
> Of where the Burst Appears; and for a Space
> List to my Calling Times while they make Dots.
> Then Hint, 'Enough!' And so we make the Pace.[41]

Milne's main form of relaxation was taking brisk walks. Striding over Portsdown, the hill above Portsmouth, he revealed his passion for mathematical logic by reciting to his assistant, Jim Crowther,[42] long passages

---

[40] Letter from Milne to G. Milne, 25 November 1917, EAM, MS.Eng.misc.b.423.
[41] Copy in the author's possession.
[42] James Gerald Crowther (1899–1983). Lecturer, broadcaster and author of more than thirty books. In World War II he was Director of the British Council's Scientific Department.

from Whitehead and Russell's *Principia Mathematica*. According to Crowther, who was to become Britain's first scientific journalist, Milne's intense, unbroken intellectual activity was almost unparalleled. Some even thought he might be another Isaac Newton.[43]

Of all Milne's adventures, flying was the most exhilarating and the least expected. In his wildest dreams as a schoolboy he cannot have imagined that his mathematics would transport him to the nose of a plane. His first flight came about by chance at Gosport Aerodrome when he happened on a school friend, George Heseltine,[44] who was training to be a pilot. He persuaded his instructor to give Milne a quick spin. So, borrowing George's leather coat, cap and goggles, Milne climbed into the shallow bucket seat in front of the pilot and off they went. The flimsy little plane, an FE2b (Farman Experimental), was a pusher plane, its propeller facing backwards behind the cockpit, no more than thirty feet in length, and with a maximum speed of 80 mph (Fig. 5). Exposed to the elements, the bumpy ride was noisy and fiendishly cold — but utterly exciting.

> I felt I was being turfed out sideways … clinging desperately to the bars …
> I couldn't for the life of me have told which way the earth would have to
> twist to come right again … Suddenly we seemed to be diving down
> straight for the sea. Quite quickly we came horizontal and then my feet
> went right up in the air … and I felt my inside protest strongly … We
> dropped to the Earth by a long spiral glide with the engine off. … For a few
> moments I was absolutely deaf.
>
> I enjoyed intensely every minute of the ride, quite possibly I shall never
> have another.[45]

Quite the contrary. Time and again he went up to measure air temperatures and pressures by leaning out over the side of the cockpit to read instruments that he had strapped to the wing. Without harness or parachute it was dangerous work, but thrilling beyond measure.

---

[43]Crowther, J.G. *Fifty Years with Science,* London: Barrie and Jenkins, 1970, p. 12.
[44]George Couleham Heseltine (1895–1980). Barrister and author.
[45]Letter from Milne to his mother, 2 March 1917, EAM, MS.Eng.misc.b.423.

Flying was a chancy business. Engines and instruments were unreliable and the meagre array of dials in the cockpit often registered faulty information. The pilot had no radio, and he needed a modicum of luck as well as skill to avoid damaging his frail wooden plane, fastened by wire and glue. The fuselage was made of ash, the propeller laminated mahogany, and the wing of spruce, covered in Irish linen.

In his letters Milne relived the thrill of his flights. In an FE2d, bigger and with a more powerful engine than an FE2b, he had to coax his pilot to go to the 'top', the height beyond which the plane could climb no further, to get the information he wanted:

> I have had the time of my life these last two days. ... Yesterday I went up for an hour and a half ... We topped 13,500 ft [4,050 m] and we bottomed 10°F of frost! Heavens but it was cold, but heavens how I enjoyed it!
>
> My pilot nose-dived several times ... It is glorious when you feel you are just going to be tippled off into space. ... The sudden change of pressure gave me rather a painful feeling in my ears.
>
> My pilots both became jocular when at the top — one started singing ... Luckily for me the engine drowned it all. It is rather boring that one cannot talk to the pilot. You have to write on scraps of paper, or contort your visage; in the latter case the pilot always misunderstands you.[46]

Soon the pusher planes were obsolete, superseded by tractor planes with their propellers facing forward. Behind the pilot for the first time in a Sopwith, Milne had a more restricted view. After a flight in an Avro he wrote:

> Last Monday I had another joy-ride de luxe. ... At the top it was minus 8½°C, fifteen degrees of frost Fahrenheit. ... It was so devilish cold that [my pilot] played for all he was worth to lose height ... he started doing a whole series of 'stalling turns' ... The machine ... starts dropping ... and for one hideous sequence of moments you feel loose in your seat.
>
> It is a weird sensation. ... After millions of these, we went into a spin. ... You feel a violent cross-wind and have to put out your arm to stop yourself leaning out over the side of the machine ... the Major of the squadron

---

[46]Letter from Milne to G. Milne, 24 May 1917, EAM, MS.Eng.misc.b.423.

congratulated my pilot on his stalling turns … apparently I have been taken up by crack fliers.[47]

As the war progressed and the death toll of soldiers mounted, conscription was extended. The government reduced physical requirements and Milne was liable for call-up along with Hawkins and William Hartree's son, Douglas,[48] now a Brigand, who pioneered techniques for solving differential equations. Hill feared that he might lose them. He could ill afford to do so and was on the lookout for more mathematicians, going to the lengths of scanning lists of Cambridge scholarships and the wounded for potential Brigands. In a ploy to keep the three, he beseeched Captain Bowring to get them commissions in the Royal Naval Volunteer Reserve (RNVR) and have them seconded to the *Excellent*. Bowring played along with Hill and ordered the sentry at the gate to arrest any predatory army recruiting sergeant who might appear unannounced! Eagerly Milne revealed the prospect to Geoffrey:

> Working the Admiralty through the Director General of Munitions Designs we should probably find ourselves within the month full lieutenants RNVR. These things have a habit of not coming off … But it will do no harm anyhow if you picture yourself saluting me, who shall be adorned as to the cuffs with a pair of curly golden bands, surmounted by a loop. The pay would automatically drop some £30 per annum … so there are compensating disadvantages.[49]

Hill had extra reason for wanting to retain Milne. He had just solved an exasperating and tiresome puzzle about a much-used formula that was consistently inaccurate because it predicted too long a range for high-angle shells, i.e. they fell 'short'. Milne identified the root of the trouble by spotting that the formula was wrong dimensionally. He introduced a term for the velocity of sound (which varies with air temperature), which made the formula trustworthy.

---

[47]Letter from Milne to G. Milne, 26 August 1917, EAM, MS.Eng.misc.b.423.
[48]Douglas Rayner Hartree (1897–1954). An expert in the science of computation, he built Britain's first differential analyser. Professor at Manchester from 1929 to 1945 and at Cambridge from 1945 to 1954.
[49]Letter from Milne to G. Milne, 8 May 1917, EAM, MS.Eng.misc.b.423.

Hill's ruse paid off and the three young Brigands were hustled into uniform — Hill remarked that Milne looked more like a schoolboy than a naval officer (Fig. 6). With a conspirator's glee Milne confided to his brother:

> Word has just come through from Richmond, who heard it from Hawkins, to whom the Commandant of Whale Island whispered it, that our commissions have come through. Yours, (RNVR) Arthur.[50]

Milne's commission, however bogus, qualified him to join the Travelling Ballistics Party, and in the summer of 1917 he went on tour with Brigadier General Drake inspecting AA defences in Scotland.

---

[50]Letter from Milne to G. Milne, 24 May 1917, EAM, MS.Eng.misc.b.423.

Chapter 4

The Trials of Trumpets

The future of aerial defence at night lies in co-operation between sound-locators, searchlights and our own fighting planes.[1]

In 1918 Milne faced a new task. 'I am to oscillate between here, Stokes Bay and Rochford,' he wrote jauntily to his father, 'carrying unto the ends of the earth the gospel of trumpets.'[2] The biblical allusion was a jesting nod to his father, for these trumpets were not the blowing kind; they were for listening. Huge, cumbersome ear trumpets could pick up the noise of a distant plane and they were the only means of locating aircraft when hidden by cloud or darkness. At night, once the position of an enemy plane was known, searchlights could illuminate it ready for gunners to take aim.

These unwieldy trumpets had existed for several years, as had 'sound mirrors', and they continued in use until radar replaced them. In 1914 blind people, deemed to have better hearing than sighted people, operated the trumpets. (Night defence of London relied on trumpet units at Lambeth Bridge, Hyde Park Corner and Charing Cross.) After sighted people were trained, the technique proved more effective, and by November 1917 thirty sets of trumpets had achieved 'very striking results'.[3]

Notwithstanding this improvement, further progress was essential because German bombers outclassed our own. The latest menace, more fearful than the Zeppelin or Gotha, was the ponderous Giant that dropped two tons of explosives. Hill asked Milne to refine the mathematics of trumpets and to liaise with the Royal Flying Corps to make them more efficient.

---

[1]Monthly Reports of the Ministry of Munitions, 1 May 1918, MUN 5/119/700/6, NA.
[2]Letter from Milne to S.A. Milne, 9 March 1918, EAM, MS.Eng.misc.b.423.
[3]Monthly Reports of the Ministry of Munitions, 12 November 1917, MUN 5/119/700/6, NA.

Basically the trumpets were artificial ears connected to the listener's own ears by rubber stethoscope tubes. Our brains interpret the sound falling on our ears to tell us its direction, and the larger the distance between the ears, the better the accuracy. (The binaural nature of hearing is comparable to the binocular property of vision.) It was this feature which the trumpets exploited. By placing trumpets five feet apart, ten times the distance between our ears, accuracy of direction increases tenfold. Operating the trumpets was an acquired skill. Horace Darwin expedited the learning process by modifying an indoor teaching simulator. He made the connecting tubes adjustable in length, so that one could be shorter than the other, which had the effect of the sound appearing to change direction. As well as speeding up instruction, the modification proved useful in assessing the competence of an operator.

A stout wooden frame housed the cumbersome apparatus, which comprised two pairs of trumpets, rotating independently about orthogonal axes (Fig. 7). One man swivelled the horizontal pair until he judged the sound in his ears to be evenly balanced, i.e. straight ahead. This gave the compass bearing of the source of the sound. A second man swung the vertical pair until the sound was equal and this gave the angle of elevation. The operators relayed their information to the third member of the team, the 'sights' man, who corrected the direction for wind etc., before passing it on to the searchlight party.

One way of improving the trumpets was to lower the threshold of audibility, the quietest noise they could 'hear'. The threshold depended on the trumpet's size and shape (determined by the angle of the cone), equivalent to the human external ear. Milne, Christopher Jakeman[4] of the NPL and Captain Archibald Ward of the Royal Flying Corps experimented with all sorts of trumpets to find an optimum design. They noticed that nearby extraneous noise tended to mask the faint hum of a distant plane, and tried to eliminate this nuisance by lining trumpets with felt, linoleum, blankets and longhaired goat skin, a material used for troop coats.

From their investigations they specified a 'standard' trumpet. This was a metal cone three feet long and two feet wide at its open end, protected

---

[4]Christopher Jakeman (1874–1959). One of the first physicists Glazebrook appointed to the newly created NPL.

by a sixteen-sided pyramidal fluted sheath, wrapped in hessian cloth secured with glue, and finished off with several coats of paint. Using such trumpets, skilled operators could locate a plane as far away as six miles to within half a degree.

Our allies, the Italians, French and Americans, accepted the superiority of the British model. The Americans had experimented with trumpets up to eighteen feet in length, convinced that 'bigger' meant 'better', but neither they nor the French incorporated corrections for wind and air conditions, which Milne devised.

Trumpets were made at Imber Court,[5] in Surrey, where Corporal (later Major) William Tucker,[6] an acoustics expert, ran a laboratory. He had already transformed sound-ranging (deducing the position of a field gun by collecting the noise of its firing on an array of microphones), into a viable technique. Like the trumpets, unwanted local noise hampered reception of the sound coming from a gun. Tucker cleverly avoided this difficulty with his hot-wire microphone, which was sensitive only to low frequencies, such as the boom of a gun.

Obviously it was imperative for trumpet operators to distinguish between the drone of an enemy plane and the drone of our own. To analyse aircraft noise, Tucker rigged up a siren at the top of a portable mast that could be wheeled about. He found that the frequencies of all planes were an octave below Middle C and that each plane had its own unique frequency.[7] The defining characteristic of a Gotha, however, was not its frequency but 'beats', readily picked up by an experienced operator.

Since it was not easy for an operator to pinpoint precisely the direction of a moving sound a 'bracketing' procedure was introduced. By swinging his pair of trumpets from side to side (or up and down) across the sound the operator set limits to the audibility, and the average gave the direction of the sound.

An important factor that had to be taken into account when using the trumpets was the 'lag of sound'. By the time the sound reaches the

---

[5] Now used by the Metropolitan Police.
[6] Willliam Sansome Tucker (1877–1955). Lecturer at Imperial College London. In 1950, he became Director of Research for Air Defence at Colborne, Ontario.
[7] The frequency of a Gotha was 80 Hz, an Avro 90 Hz, an FE2b 70 Hz.

trumpets on the ground the plane has moved on, by as much as half a mile. This discrepancy depends on the speed of sound, which varies with temperature. (Sound travels more slowly at lower temperatures.) The correction for the 'lag of sound' was addressed by introducing a special ring-sight attached to the gun.

Further improvement came with a more sophisticated method of communicating between the trumpet team and the searchlight team, some ten yards distant. Instead of the 'sights' man shouting out the readings for elevation and compass bearing, which took time and might be indistinct, he selected switches on a silent indicator that displayed appropriate words.

There were no headquarters for trumpets, and Milne was constantly on the move, sometimes away from HMS *Excellent* for weeks at a stretch. He relied on his bicycle for getting to such places as Stokes Bay, near Gosport, where operators were trained and where the Royal Engineers' School of Electric Lighting carried out trumpet trials with searchlights. Even if he went by train he often took his bicycle in order to pedal the remaining few miles, for instance from Beaulieu Railway Station to Beaulieu Airport. Once, when he thought he had lost his machine, he was distraught.

Travelling between these sites, Milne encountered the food shortages that were commonplace to civilians. Despite being expected at military establishments, his fat and sugar rations sometimes failed to materialise. Passing through London he saw that restaurants were restricted to fish, egg and vegetarian dishes. The government exhorted everybody to eat less meat and to keep chickens, as Milne's parents did.

While Milne was working with the Royal Flying Corps at Rochford Aerodrome in Essex, he lived in a boarding house at Prittlewell, on the outskirts of Southend, in a converted railway carriage with neat white furnishings, parked in the garden. The intention was to carry out trials at night, but Rochford possessed hardly any planes equipped to fly in the dark, and the trumpet teams 'relied mainly on the enemy to provide the targets'.[8]

---

[8]Letter from T.L. Wren to H.F. Baker, 13 January 1929, St John's College Library, Cambridge, Miscellaneous Papers.

It transpired that identical trials were in progress at the RFC's Experimental Station at Orfordness,[9] in Suffolk, whose intrepid commander was Colonel Bertram Hopkinson,[10] Cambridge's Professor of Mechanical Engineering, who learnt to fly the better to carry out his job. To save duplication, Hill wanted Milne to conduct trumpet trials at Orfordness but the military obstructed his transfer. Once again personal relations saved the day. Hill was well acquainted with Hopkinson and asked him to intervene. 'A few wise decisive words spoken at the critical moment'[11] worked the trick, and Milne took the trumpets and four men to the station.

The Ness is a low-lying bleak peninsula, reached by ferry across the River Ore. German prisoners maintained the seawalls which protected the 10-mile windswept stretch of shingle from being pounded to bits by the North Sea. It was there that Hopkinson led a courageous team who investigated unexploded bombs and dangerous, often outlandish, aeronautical gadgets.[12] In this stimulating environment, Milne enjoyed 'amusing incidents'.[13] He passed one lively evening with the Trinity mathematician Geoffrey Taylor,[14] then a second lieutenant, who flew in from the Royal Aircraft Establishment, Farnborough, where he was a member of the famous Chudleigh Mess.[15]

Milne stayed for two months and his work flourished. According to an official report, 'The system being developed at Orfordness now is likely to prove by far the most effective means of dealing with night raids.'[16]

---

[9] The hangars and sheds were later used for the secret development of radar. Today the nature reserve on the Ness is owned by the National Trust.

[10] Bertram Hopkinson (1874–1918). Patent lawyer and engineer. He was elected Professor of Engineering at Cambridge in 1903.

[11] Hill, A.V. *Memories and Reflections*, p.235, AVHL I.

[12] For example the Guardian Angel parachute slung beneath the plane. The pilot had to clamber onto the wing, grab the parachute, put it on and jump clear before the plane tail clipped him.

[13] Letter from Milne to S.A. Milne, 9 March 1918, EAM, MS.Eng.misc.b.423.

[14] Sir Geoffrey Ingram Taylor (1886–1975). Royal Society Yarrow Research Professor from 1923 to 1952.

[15] The Chudleigh Mess included three future Nobel Prize winners: E.D. Adrian, F.W. Aston and G.P. Thomson. F.A. Lindemann, who identified how to counteract 'spin' and verified his theory by courageously going into a spin on a solo flight, was also part of the group.

[16] Monthly Reports of the Ministry of Munitions, 1 March 1918, MUN 5 119/700/6, NA.

Milne's heart was in his work and he was confident of its worth. He deplored the public resignation to German raids and scorned the stuffiness of the military:

> It is said Londoners know that such incidents are inevitable ... This is complete bilge. ... I should like to say that the methods developed by the RFC at Orfordness for night attack on Gothas, combined with their location by the methods developed by MID and RFC (my work at Orfordness etc.) provide an absolutely reliable way of bringing down all Gothas entering a defended area. But the generals won't believe it. ...
>
> Unfortunately generals have no imagination ... and there is jealousy between gunners and the RFC. ... Just because the new methods cannot be patched on to the old, but require a completely fresh start ... nobody with crossed swords and a pip on his shoulder strap will go for them.[17]

Milne's biggest contribution was to refine the mathematics describing the path of sound as it travels from a great height down to the trumpet operator. The path is curved, the bending being akin to the refraction of light through water, but with one complicating difference. As the sound gets closer to the Earth the air becomes both denser and warmer, so that the refraction is always changing. It is this which causes the curvature. Furthermore, when the listener finally hears the sound it is distorted, comparable to the mottled appearance of an object viewed through ribbed glass.

Another key factor was the wind. From his previous work on shells, Milne had a head start in investigating its influence on the passage of sound. According to Tucker, Milne's insight broke new ground and added to the classical studies of the nineteenth century. 'The range of audibility and meteorological conditions was worked out by Stokes, Reynolds, Rayleigh and Milne in terms of the velocity of sound with height above the Earth's surface, the velocity being affected by wind and temperature.'[18]

---

[17]Letter from Milne to S. Milne, 9 March 1918, EAM, MS.Eng.misc.b.423.

[18]Address given by W.S. Tucker on 9 September 1930 at the British Association for the Advancement of Science in Bristol. *British Association for the Advancement of Science: Report of the ninety-eighth meeting, Bristol, 1930, 3–10 September*, London: British Association, 1931, p. 306.

Milne was fulfilled in his work but grew melancholy. The war was at stalemate and as it dragged on into its fourth year he despaired that it might never end. The mood of the country had plummeted from high patriotic fervour to cynical dejection. Besides, he was worried about his father, who was suffering from a weak heart. On doctor's orders he stayed at home. Away from his school, he grew depressed and his pay was docked. He protested, and, eventually, backed by the National Union of Teachers in a case that set a precedent, he secured a refund. A further cause of stress was that Sidney Milne, aged fifty, was under threat of conscription, as the 1918 Military Services Act applied to men up to the age of fifty-one. Milne was indignant:

> I feel so strongly about the ridiculousness of calling up older men when they make such poor use of the men they have got, that the only thing I can say about it is 'Balls'. The way men are wasted ... makes one weep.[19]

Up and down the country families suffered bereavement and the Milnes were no exception. Uncle Fred, a Master Mariner in the merchant navy, went down with his ship in the North Sea. Milne's favourite cousin, Clifford, a second lieutenant in the North Lancashire Regiment, was killed in action at Givenchy and posthumously awarded a Military Cross for his bravery. Milne felt his death keenly.

Geoffrey was in France. Determined to evade the censor, he managed to reveal to his parents that he was stationed near St Quentin by directing them to a certain bookshelf, where they found *Quentin Durward*. Geoffrey was in a sound-ranging unit and commiserated with his brother over dodgy field telephones. This work kept him back from the front line by several thousand yards, but in March, although his section had no combat training, they were abruptly ordered into battle by Brigadier-General Sandeman Carey[20] and assigned to a 'scratch' battalion, known as Carey's force. Its remit was to prevent the Germans breaking through Amiens and reaching Calais. Geoffrey waded through swamps, along tracks turned to

---

[19]Letter from Milne to S.A. Milne, 19 June 1918, EAM, MS.Eng.misc.b.423.
[20]Major-General George Glas Sandeman Carey (1867–1948) commanded the 20th division of the Royal Artillery.

quagmires, and lived in cellars and stables. Under fire and on an empty stomach he was exposed to shells and shrapnel. Eating a mess tin of hot stew and having a wash were precious luxuries:

> We have had a nomad's life, marching more kilometres than I care to think about ... turning out at all hours of the night ... We had a pretty hot time of it ... crossing the open with Jerry's machine guns pegging away umpteen to the dozen. However, I came through OK, thank God. Not all the chaps did.
>
> Maundy Thursday, Good Friday, yesterday, seemed to contain enough events to fill weeks, but Easter Sunday finds us safe here away from the line at H.Q. I feel like sleeping for three days on end. ... I didn't know what fighting, or rather killing, was, until then, but I think three days have given me my fill of it.[21]

Geoffrey's letters and field cards wracked Milne with guilt. He felt compelled to take a more active part in the war and Clifford's death brought him to a decision. He asked Hill to release him, aware that Hill had recruited some wounded officers as mathematicians. Among them were Lieutenant Charles Mayes,[22] who wrote a poem about the Brigands in the style of *The Song of Hiawatha,* and Lieutenant William Dean,[23] a Trinity friend, whose calm manner Milne valued in moments of stress.

Milne admitted that his work was 'as interesting as anything I could wish for', but 'is it fair that I should be here in England, sheltered, and doing highly congenial work, at this stage of the war?'[24] Hill's reply was unambiguous:

> I may be quite wrong-headed about the importance of brains, but personally I should most strongly advise you that your duty to your country is to remain where you are ... I feel that I, as an able-bodied man, ought to be fighting, but I am convinced that I am more likely to be useful in my present capacity than in any other ... The same is true for you.[25]

---

[21]Letter from G. Milne to S.A. Milne, 31 March 1918, in the author's possession.
[22]Charles Mayes (1897–1970) became a housemaster and head of science at Eton.
[23]William Reginald Dean (1896–1973). Goldsmid Professor at University College London from 1952 to 1964.
[24]Letter from Milne to Hill, 26 May 1918, in the author's possession.
[25] Letter from Hill to Milne, 27 May 1918, in the author's possession.

Milne persevered with the trumpets. He organised elaborate trials with trumpet-searchlight teams, placed three miles apart, with the object of getting both sets of searchlights to illuminate a plane simultaneously, thus making it easier for the gunners to see. A 'raider' plane, usually a Bristol Fighter, flew as a target. Once it was illuminated by the two searchlight beams, the 'chaser' plane, typically a BE12, went after the target and lined it up in its gun-sights. A good performance was rated as six sightings in forty minutes, the 'chaser' having deliberately lost the 'raider' between sightings. To a great extent, obtaining simultaneity hinged on the wretched field telephones. After trials near Rochford, Milne brought his teams to the south coast, first to Chichester and Selsey, then along the stretch between Portsmouth and Southampton. He worked long hours:

> Last Sunday ... I slogged through my report, and was off to Chichester with Wren[26] by 9 pm. As usual we got to bed at 2 am and I was up again at 7 am to get away to London.[27]

A month later:

> Sunday was rather strenuous, shooting out at Eastney until 2 pm (no lunch), and then a trip to Chichester and beyond in the evening, returning at 1:30 am ... I go to Imber Court next Monday, probably for a week.[28]

Milne loathed the army's petty bureaucracy, such as signing chits, and bemoaned his fate as officer in charge at Rochford when Captain Ward left to take a load of trumpets to France. In August it was Milne's turn to go. He was granted embarkation leave, and issued with a Journey Meal Card which he exchanged for rations at his Food Office in Hessle. It was usual practice for navy personnel on field duties to wear khaki to blend into the countryside, and Milne got himself fitted out in a smart new uniform. He wore it for the photograph he gave his parents, the usual present of departing servicemen.

---

[26]Thomas Lancaster Wren (1889–1972). Mathematician. Milne had attended his lectures. In 1919 Wren was appointed a reader at University College London.
[27]Letter from Milne to S.A. Milne, 19 June 1918, EAM, MS.Eng.misc.b.423.
[28]Letter from Milne to S.A. Milne, 25 July 1918, EAM, MS.Eng.misc.b.423.

Milne crossed the Channel with a consignment of trumpets and his Trinity friend Lieutenant Sam Pollard, now a Brigand. Their destination was the headquarters of British Anti-Aircraft Defence at Divion, ten miles from Saint Pol. After reporting to Colonel Douglas Gill they settled into the sound location centre a few miles away at Camblain where they instructed men on how to use the trumpets.

> The Colonel has been down to see us two or three times, and has been most interested [in our work]. ... Ever since we started, however, the weather has been atrocious. Until this afternoon it has hardly stopped raining for an hour at a time and it has been blowing 20 miles per hour ...
>
> I said men talked a lot about leave, well, they do — it is the thing most looked forward to. Though they are supposed to get it every 5 months in the batteries here, the time is averaging 7 and 8 and more months. ... as for whisky, well for 5 people out of 6 ... a whisky and water is the staple drink at dinner .[29]

In the evenings they relied mainly on candles, and even though Milne managed to buy a torch he had it pilfered from his tent by French children. Yet this was a minor peccadillo compared with British looting. In another letter, written in pencil:

> I have just finished a detailed report of my work ... It has been a long business as I have negotiated all the arithmetic myself. ...
>
> Telephone lines continue to give me unfailing trouble. They eternally break, or wear out, or get cut by Frenchmen (unless the French cows go about with wire cutters tied to their heels and hoofs). Or the speaking and listening part goes wrong. Moreover they always get found out only at 11 pm and you have to go haring round in the dark. ...
>
> The word 'loot' never actually appears in our newspapers ... Actually it is the key to a comfortable life out here. Most of the chairs in the mess and nearly all the pots and glasses are loot from a nearby French town ... The O.C. of the School here rarely goes off on a jaunt without collecting a chair or so.[30]

---

[29]Letter from Milne to S.A. Milne, 11 September 1918, EAM, MS.Eng.misc.b.423.
[30]Letter from Milne to S.A. Milne, 22 September 1918, EAM, MS.Eng.misc.b.423.

On 11 October, Milne entertained the distinguished engineer Sir Alexander Kennedy during his inspection of AA defences. A staunch friend to the Brigands, he had attended Hill's trials at Northolt and Bushy Park and was familiar with Milne's contribution. Kennedy reported that he was 'very pleased with the ingenious work'[31] done by Milne at Camblain.

While at Arras, a city riddled with underground tunnels, Milne was ordered to Paris to advise the Americans on trumpets. Paris was only a hundred miles away, but with tortuously slow trains and a circuitous route, his journey lasted three days. On 2 November he hitched a lift in a lorry back to Divion to collect rations before catching a train to Etaples where he had a comfortable night in the adjutant's quarters. He spent the following night in Rouen on a chair in a crowded hotel. Once in Paris, he spent his mornings at the French Ministry of Inventions, which left him with plenty of free time to explore the city, 'money dribbling out of my pockets the whole time'.[32] On the historic morning of 11 November he was at Saint Cyr, the French military academy near Versailles, when he heard the glorious news of the armistice. At last the fighting was over, and with an enormous sense of relief all work immediately ceased. He saw an old cannon fired from the Grande Place at Versailles in celebration, and that evening he was on the heady streets of Paris.

A week later Milne was on his way to England for a three-day leave. At Hessle he found his father far from well. Geoffrey was still in France, in hospital, having succumbed to the 'flu' epidemic. He was home by Christmas when Milne had ten days leave and they had a happy reunion, swapping stories and poring over high scale maps of battle areas that Milne had procured.

In the elation of the peace and released from wartime pressures, the Brigands indulged in some frivolous nonsense. The hundred people who had 'worked with us, helped us, or aided and abetted us'[33] sealed their camaraderie by binding themselves into 'The Honourable Society of Brigands'. This was ratified by 'A License to Practise as a Brigand', a decree written in copperplate on an enormous scroll that combines, in

---

[31] 'Notes on a Visit to France and the First Army' by A. Kennedy, 9–20 October 1918, AVHL I 3/39.
[32] Letter from Milne to G. Milne, 19 November 1918, in the author's possession.
[33] Hill, A.V. *Memories and Reflections*, p. 511, AVHL I.

flowery language, the citation for an OBE, a commission in the army, and the Home Office permit for experimenting on living animals. Mocking pompous officialdom, it begins:

> We A.V. Hill by the grace of Mathematics Physics Chemistry Phudgiology of the United Anti-Aircraft Experimental Section of Portsmouth Rochford London Stokes Bay and Cambridge, Director and Principal of the most Excellent Respectable Learned Temperate Assiduous and Ancient Society of Brigands, to our Trusty and Well-Beloved, Greetings ...[34]

Hill, accorded the position of 'One of His Majesty's Principal Brigands', headed the list of satirical titles, such as 'Arch Spinner', 'Principal Pickpocket' and 'Chief Noise Producer'. Milne was an 'Unprincipled Brigand'.

The license was illustrated by a cartoon and solemnised in 'The Ballistic Faith', an irreverent parody of the Athanasian Creed, inspired by a notice at Woolwich that compared three artillery divisions (horse, field and heavy) to the Holy Trinity. 'The Ballistic Faith' begins:

> Whosoever will be appointed a Brigand; before all things it is necessary that he hold the Ballistic Faith.
> And the Ballistic Faith is this; that we worship one mirror position finder in three stations, and three stations in one mirror position finder.[35]

The blasphemy upset one of the older Brigands,[36] but Milne went along with it. His experience of working seven days a week had eroded the sanctity of Sunday and his Christian beliefs were wobbling. Later, after he emerged from a period of agnosticism, he expressed scruples about the 'Faith'.

The cartoon, captioned 'Ante-Aircraft Gunnery', depicts a fiery winged dragon gliding unharmed above a ferocious medieval battlefield. Milne framed his copy and hung it in his study. If it drew comment he glowed with pleasure as he reminisced about his extraordinary Brigand adventures.

---

[34] A copy of the License is in AVHL I 1/35.
[35] Copies of the cartoon and Ballistic Faith are in AVHL I 1/35.
[36] Charles Edward Haselfoot (1864–1936). A rare Oxford Brigand. He became Senior Tutor, Bursar and Dean of Hertford College.

Public recognition of the Brigands' outstanding achievements came with the award of OBEs to Hill, Fowler and William Hartree, and an MBE to Milne. The Brigands expressed their gratitude to Captain Bowring and his officers with the gift of a specially commissioned pendulum clock, its design so intricate that it is all but impossible to tell the time from it. When I saw it, it occupied a prominent corner of the wardroom at HMS *Excellent*. Below the inscription on its brass plate that records 'unvarying kindness' for 'happy strenuous days', are the names of twenty-five Brigands, listed in the order of joining. Milne's name is third.

Milne was perturbed about his future. Most of his contemporaries, Crowther, Dean, Douglas Hartree, Hawkins and Mayes, intended to finish their degrees at Cambridge. Milne was strongly tempted to do so too, because he dearly wished to continue in mathematics, but he was troubled about allowing himself the luxury of further study when his father was ailing. If Milne were employed the family would be sure of having at least one breadwinner. Since 1914 the pound had halved in value, the pinch of inflation threatened, and his parents, always prone to worry, feared poverty. Milne decided that he must find a job.

As soon as Hill heard of Milne's predicament and his reasons for seeking a job, he took prompt action. He increased Milne's pay to £350 per annum and back-dated it for two months so that he was able to send money home. As for the idea of a job, Hill dismissed it out of hand. He was sufficiently impressed by Milne's mathematical ability to be confident that he could forge his way in academic life. He was adamant that Milne abandon any thoughts of further undergraduate study and audaciously recommended him to aim directly for a prestigious fellowship at Trinity College. This had a double advantage. It would give Milne the freedom to pursue mathematics and enough financial independence to be able to help his family. Hill wrote to Milne:

> I am quite determined that you are going to return to Trinity, so don't think you are going away elsewhere to 'start earning' … I imagine you are very nearly a certainty for a fellowship in 1919 or 1920. Herman, Fowler and I — not to mention Richmond — can speak sufficiently strongly of your capabilities to make it pretty likely they will elect you in 1919 or 1920, and then with £250, plus room and dinners plus commons, approximately equal

to £350, plus what you can make by teaching, you will be actually better off than by going away somewhere else — and what is more you will be doing what it is your right and duty to do — to help in the promotion of knowledge.

If you like, and will give me a fairly detailed account of what your private situation is as regards money and prospects at home, I will immediately see [R.V.] Laurence and the Master about you and see what they advise.[37]

True to his word, Hill approached Sir Joseph Thomson, who had succeeded Dr Butler as Master, telling him of an 'extremely able and estimable member of Trinity College, E.A. Milne, now a Lieutenant RNVR, who has been with me for 2½ years'.[38]

Milne, ever given to caution, was wary of the gamble. Along with other fellowship candidates he would be judged on the merits of papers, as yet unwritten. He questioned the prudence of risking the months he needed to prepare these papers, for, if he failed to win a fellowship, he would have no job and all the best ones would have been snapped up. To allay Milne's qualms, Hill arranged for him to talk to Sir Alexander Kennedy. He was old enough to be Milne's grandfather and had a persuasive knack with young men. His arguments tipped the scales and Milne made the crucial decision to try for a fellowship. It proved to be the turning point of his life.

Meanwhile, while winding up his MID work at Whale Island, which entailed more blustery trials and tedious reports, Milne paid two fleeting visits to Cambridge. G.H. Hardy pointed out that Milne would be the most senior young mathematician returning to the college and charged him with setting up a mathematical society. On consecutive Saturdays in February Milne went to Cambridge, and in Sydney Chapman's rooms changed from his RNVR uniform into civilian clothes. Milne canvassed for members, drafted the constitution and rules of the Trinity Mathematical Society[39]

---

[37]Letter from A.V. Hill to Milne, 25 November 1918, in the author's possession.
[38]Letter from A.V. Hill to J.J. Thomson, 29 November 1918, Thomson papers, Cambridge University Library.
[39]The Trinity Mathematical Society predates the Adams Society of St John's, founded in 1923, and the Archimedeans, the university mathematical society, founded in 1935.

and set up a preliminary meeting. He was demobbed in March, moved into rooms at Trinity in April and gave the society's inaugural paper in May. He tenderly nurtured the infant society, and nervously harboured misgivings that it might not survive the initial show of enthusiasm. Happily it took root and it flourishes to this day. J.J. Thomson spoke at the 100th meeting, Rutherford at the 150th and Milne at the 200th.

Milne's Brigand years sparkled with excitement. He was thrown into stimulating company, travelled widely and a wealth of new experiences broadened his view of the world. More importantly, the opportunities he was given in ballistics unleashed his extraordinary capacity for original thinking and converted him into a mathematical physicist. Milne never forgot his huge debt to Hill, nor his far-reaching influence: 'In common with the other Brigands I owe a good fraction of what has happened [to me] to your Chief Brigandship'.[40] As Hill succinctly put it, 'Kaiser Wilhelm and I, jointly, did a good service in science in diverting both R.H. Fowler and E.A. Milne from pure mathematics to other fields'.[41]

---

[40]Letter from Milne to A.V. Hill, 20 February 1926, AVHL I.
[41]Hill, A.V. *Memories and Reflections*, p. 41, AVHL I.

Chapter 5

Cambridge Rhapsody

Here was friendship; here was freedom in fullest abundance.[1]

After being elected a Fellow of Trinity, Milne wrote euphorically, 'Friday was certainly the greatest day of my life'.[2] During the anxious months of preparing his fellowship papers he had watched his expenses carefully. He had revived part of his Yorkshire scholarship money and Trinity had tided him over the summer with a Jeston Exhibition of £50 and a Yates Prize for 'impoverished' students. Now he exhilarated in the knowledge that his gamble had paid off and his future was secure. What he did not foresee was that the fellowship would propel him into the orbit of the most celebrated physicists and mathematicians of the day and accelerate his entry into professional bodies. Just two months later, in December 1919, he was elected to the London Mathematical Society.

The largesse of the fellowship exceeded his wildest dreams. He had stepped through a magic door into a utopian world and he marvelled at its string of privileges. Beyond his food and accommodation he could take guests to dine at the High Table. He no longer had to keep to the paths in the Great Court but could walk across its cropped lawns from his ground-floor rooms opposite the Chapel. With a college key in his pocket he could come and go as he pleased, regardless of when the porter locked the main gate, and enter hidden sanctuaries, the Fellows' garden and bowling green. Life was sweet.

After the fellowship admission ceremony in the Chapel, Hill took Milne out to lunch and tipped him off that he should wear a dinner jacket to hall that evening. Dreading a dull stuffy occasion, he was pleasantly surprised by the warmth shown to him. He felt honoured to be seated next

---

[1]Trevelyan, G.M. *An Autobiography*, London: Longmans, Green, 1949, p. 13.
[2]Letter from Milne to his parents, 12 October 1919, EAM, MS.Eng.misc.b.423.

61

to the Master, Sir Joseph Thomson, President of the Royal Society, who, Milne noticed, ended his sentences with a resonant half-grunt laugh. On Milne's other side was the historian Denys Winstanley who had a ready flow of conversation. After dinner Thomson entertained the Fellows to a smoke at the Master's lodge. (Until the harmful effect of nicotine was discovered, 'having a smoke' was the understood gesture of masculine friendship.) Then they went on to G.H. Hardy's rooms and it was past midnight before the party dispersed.

Milne rated the fellowship the highest peak of achievement. In his case it was an unusual feat, too, because he had not completed his degree, and it testified to the sophistication of his Brigand mathematics. The four papers he submitted, polished up with points of style instilled by G.H. Hardy, were all published. The paper of greatest import was about the upper air, on which he collaborated with Sidney Chapman, an authority on ionisation in the Earth's atmosphere. Chapman wanted to collate and scrutinise the mass of data that Milne had obtained by hanging out of aircraft. In their monumental forty-page paper they organised the data into tables which, amongst other things, compared the amounts of various gases at various heights. This suggested that there was no clear demarcation between the troposphere and the stratosphere, an idea which ran counter to prevailing wisdom. In June 1920 Milne had his first experience of giving a scientific paper to a learned body when he put their revolutionary ideas to a meeting of the Royal Meteorological Society, chaired by Sir Napier Shaw,[3] the Director of the Meteorological Office since 1905. The landmark paper changed opinion. But Chapman and Milne were extraordinarily lucky because by a happy fluke the four mistakes in their calculations cancelled each other out.[4]

Of his other three fellowship papers, one on sound waves was translated into French for the benefit of their artillery. He and Sam Pollard, also a successful fellowship candidate, composed a paper on errors relating to

---

[3] Sir William Napier Shaw (1854–1945). Expert on the upper air. Author of the reference book *Manual of Meteorology*, Cambridge: Cambridge University Press, 1919. He introduced the millibar unit of atmospheric pressure.

[4] King-Hele, D.G. 'The Earth's Atmosphere; Ideas Old and New', in Bondi, H. and Weston-Smith, M., (eds), *The Universe Unfolding*, Oxford: Clarendon Press, 1998, pp. 145–147. (M. Weston-Smith is the author's daughter.)

sound trumpets. The fourth was to be on polynomials, but at the last minute Milne withheld that paper because he doubted its worth and substituted one of his MID reports on the geometry of sound trumpets.

During these months Milne was also busy on a confidential War Office textbook on sound location that remained valid for years to come. (Richmond edited companion volumes on AA gunnery.) He was not enamoured by the task but could scarcely refuse when Hill asked him to do it. Milne co-ordinated contributions from Jakeman, Tucker and Ward, chose illustrations and diagrams, and wrote a chapter. The book covers the history and principles of sound location, its equipment and the psychological and practical difficulties encountered.

In 1919 Cambridge was recovering from the disruptions of the war and was flooded with ex-servicemen. The university was mindful of the need to modernise, with telephones and typewriters, but still fostered hallowed traditions like locking up undergraduates inside college at night and requiring gowns to be worn on Sundays, even on a country walk. It was not surprising that ex-servicemen, who had survived the mortal ordeals of the trenches, chafed at petty footling rules; they wanted to complete their studies as quickly as possible and find employment. Conscious of their plight, the university introduced examinations in military subjects and a shortened BA degree. However, the university exempted Milne from these regulations, excusing him any further examinations and awarding him a BA degree on the strength of his Brigand mathematics.

Yet Milne paid a price for leapfrogging half his undergraduate studies. That he had not gone through the proper mill gave him a deep sense of inferiority. 'I can no more use quaternions than fly,'[5] he wrote. In spite of Hardy's encouragement, he persistently underrated himself, as a 'parish pump mathematician'.[6] When he had to do his share of examining he felt a sham, and had to swot up on topics in the syllabus about which he was ignorant.

If Milne was a Fellow in name, he was an undergraduate in spirit, free of responsibility and eager to seize opportunities to learn. Bubbling

---

[5]Letter from Milne to G.J. Whitrow, 4 October 1949, GJW, private access granted by GRW, contact Imperial College Archives.
[6]Letter from F. Smithies to the author, 26 February 1991.

with vivacity he threw himself full tilt in all directions. He rushed from one lecture to another. He heard the Dutch physicist Peter Debye[7] on the Bohr atom, John Maynard Keynes[8] on the economic consequences of peace, and attended a course on special relativity. Milne taught himself German, essential for any physicist worth his salt, so that he could read German periodicals. After buckling down to rudimentary grammar and reading simple stories, he tackled a paper by Max Planck[9] on quantum theory and made word lists until he mastered the vocabulary.

He turned down several requests for teaching but agreed to act temporarily as Director of Studies in mathematics at a small college, Trinity Hall, in order to look after its dozen undergraduates studying mathematics. Twice he gave supervisions to students at Girton, then exclusively a women's college.

He enjoyed his new-found leisure. He played some indifferent tennis and fives and won his 4th board matches at chess for Trinity. Socialising revolved around male clubs and societies, revived from their wartime abeyance. On Friday evenings Milne indulged his love of debating at Trinity's fashionable society, the Magpie and Stump, where wit took precedence over substance. Once a term, on punishment of a silver fine, each member had to speak for at least two minutes. Milne had no trouble complying with this rule and debated such fanciful motions as 'This house regrets the indiscretion of Columbus in 1492'.

The long-established Magpie and Stump affected umbrage at the upstart Trinity Mathematical Society and challenged it to a hoop race through the streets of Cambridge ending at the anti-penultimate E in Magdalene Street. The jape generated quantities of mock-serious argument about the eligibility of people like Milne who belonged to both societies. The Magpie and Stump defeated all amendments put forward by the Mathematical Society and Milne felt that the society must assert its

---

[7]Peter Joseph William Debye (1884–1966). Chemist. He held appointments in Zurich, Utrecht, Göttingen, Leipzig and Berlin. Nobel Laureate 1936.

[8]John Maynard Keynes, Baron Keynes of Tilton, (1883–1946). Bursar of King's College, Cambridge, from 1919 to 1946. He financed and built the Arts Theatre, Cambridge.

[9]Max Karl Ernst Ludwig Planck (1858–1947). Professor at the University of Berlin from 1892 to 1926. Nobel Laureate 1918.

dignity and tweak the nose of the Magpie and Stump. Sadly history does not reveal whether the race took place.

Milne joined a play-reading group, the Philistines, at the invitation of an undergraduate, Geoffrey Turberville,[10] who had rooms above him on L staircase. The group relied on the public library for copies of the chosen plays. If there were not enough to go round Milne bought his own and built up a treasured collection of Ibsen, Galsworthy, Yeats and Shaw, as well as Gilbert Murray's translations of Greek plays.

The term sped by, studded with highlights. On the first anniversary of the armistice, with memories still raw, a fresh fall of snow brought an eerie hush to the city. Milne attended an emotional service in Trinity Chapel, where two laurel wreaths draped in purple ribbon hung on either side of the altar and a bugler sounded the 'Last Post' from the antechapel. Richmond took him to a college feast at King's, his other guest being Colonel Gill. Touched that his headmaster Mr Gore came to Cambridge, Milne proudly exercised his privilege of taking him to dine at the High Table, along with Richmond, who was an undergraduate contemporary of Gore's at King's. Milne bestirred himself to entertain Mr Gore and convened the ten Hymerians up at Cambridge for a 'squash', an informal get-together in his rooms. Its success prompted him to instigate annual Old Hymerian dinners. On these occasions his gift for apt quotations was a revelation to those who did not realise he was literate as well as numerate.

Hardy (Fig. 8), who was fast becoming more of a friend than a teacher, introduced Milne to 'vint', an elaborate Russian version of bridge. In a trial game Milne managed to pass muster, his brain reeling from the drawn-out bidding as he pitted his wits against Hardy's select little 'league'. They kept running totals of their scores, with Hardy well ahead of the rest. Milne rubbed shoulders with the regular players: Ralph Fowler; the reserved and quiet atomic physicist Francis Aston,[11] who was constructing his mass spectrograph which confirmed the existence of isotopes; and Horace Darwin's nephew, Charles G.

---

[10]Geoffrey Turberville (1899–1994). Headmaster of Eltham College, London.
[11]Sir Francis William Aston (1877–1945). Research Fellow at Trinity College. Nobel Laureate 1922.

Darwin,[12] who was large and jovial, and an early pioneer in the theory of X-ray diffraction.

Throughout the term Hardy was determined to whip up support to re-instate Bertrand Russell as a college lecturer after the council dismissed him in 1916. Its treatment outraged Hardy, as it did many Fellows, especially the younger ones. Hardy's campaign bore fruit, and in December 1919 the majority of the Fellows, including Milne, voted to re-instate Russell. The college offered him a lectureship, which he accepted but never took up because after a year's leave of absence in China he resigned voluntarily. (In 1944 the council elected him to a fellowship.)

Notwithstanding this dizzy whirl, Milne was conscious of a drawback to his fellowship. At twenty-three, naïve and gauche, he was the youngest Fellow and belonged neither to the social nor the intellectual aristocracy of Cambridge. At dinner, despite a few familiar faces at High Table, he cast a wistful eye to where he used to sit, yearning to be back among his peers. He felt subdued in the presence of his august seniors — it was claimed that Trinity's sixty Fellows could supply expertise on any subject. Among them was the philosopher John McTaggart,[13] whose lectures Milne sampled. McTaggart drew a large audience, enough to fill two adjacent rooms, and he stood midway between the doorways, discoursing on the nature of time, the 'before' and 'after' of events. There was the acerbic Professor of Latin, A.E. Housman,[14] who wore a schoolboy's cap, Milne observed, when walking in the country. Written in the year Milne was born and a poem he knew well, 'A Shropshire Lad' was at the peak of its popularity. Yet Housman would not tolerate anybody mentioning it and nobody seemed to know what inspired him to compose it.

The Master, for all his distinction, was not intimidating. Unlike his fastidious predecessor, Montague Butler, J.J. Thomson was unpretentious,

---

[12] Sir Charles Galton Darwin (1887–1962). Tait Professor at Edinburgh from 1922 to 1936. Director of the National Physical Laboratory, Teddington, from 1938 to 1949.

[13] John McTaggart Ellis McTaggart (1866–1925). Philosopher. Lecturer at Trinity College from 1897 to 1923.

[14] Alfred Edward Housman (1859–1936). Professor of Latin at University College London from 1892 to 1911 and at Cambridge from 1911 to 1936.

approachable and blissfully unaware of his eccentric appearance as he entered the Chapel with his clothes awry. Ernest Rutherford, hearty and boisterous, impressed Milne with his kindness. He urged him to try his hand at some practical physics, and arranged for him to spend three mornings a week at the Cavendish Laboratory under Edward Appleton,[15] who gives his name to the layer of the atmosphere which reflects radio waves, and Henry Thirkill,[16] a more sympathetic instructor known as 'Thirks'. Manual dexterity was never Milne's forte. His heart was in theoretical physics but Rutherford would have no truck with mathematics for its own sake, bellowing out impatiently, 'What are your physical assumptions?'[17] Insisting on the highest standards, he advised Milne, 'There is no page or paragraph, or paper in a journal, which one cannot improve if one tries.'[18]

The fellowship landed Milne in sartorial embarrassment. At Trinity's Feast of All Saints, he was acutely chagrined to notice that he was almost the only person in a dinner jacket. Clearly tails were expected. This marred his enjoyment of the evening's magnificent pageantry: the red carpet in the hall, the ceremony of the loving cup, the tall figure of Alan Gray[19] in his voluminous cherry-coloured gown conducting the choir in the minstrels' gallery. Milne immediately ordered a tailcoat but it was not ready for an even grander dinner when the top brass of the war, Lord Byng, General Robertson and Sir Hugh Trenchard were to dine at Trinity after receiving honorary degrees. Milne could not face a repetition of his embarrassment and avoided the dinner.

The outstanding event of the term, indeed of the scientific year, was the talk by the great astronomer Arthur Eddington[20] about his eclipse expedition to Principe, an island off the coast of West Africa. The lecture was open to all members of the university and the queue stretched right across

---

[15]Sir Edward Victor Appleton (1892–1965). Vice-Chancellor of Edinburgh University from 1949 to 1965. Nobel Laureate 1947.
[16]Sir Henry Thirkill (1886–1971). Master of Clare College from 1939 to 1958.
[17]Letter from Milne to A.G. Walker, 12 June 1943, AGW.
[18]Eve, A.S. *Rutherford*, Cambridge: Cambridge University Press, 1939, p. 237. While Eve was researching this book in Manchester he stayed with Milne's in-laws, the Fiddeses.
[19]Alan Gray (1855–1935). Organist at Trinity College from 1893 to 1930.
[20]Sir Arthur Stanley Eddington (1882–1944). Plumian Professor at Cambridge from 1913 to 1944.

Trinity's Great Court. The hall was packed to bursting, with people sitting on the floor and on tables, and squeezing into the oriels and gallery. During an eclipse, when the Moon blocks sunlight from coming to Earth, we can see stars close to the Sun that are not usually visible. Meticulous measurements of photographs taken during the eclipse revealed the changed directions to certain stars observed near the solar limb. Eddington explained that, on balance, the results confirmed Einstein's prediction that gravity bends light and also emphasised that space–time differs from flat Euclidean space–time in the neighbourhood of a strong gravitational force, such as the Sun. In thanking him, J.J. Thomson wittily conjectured, 'Morality varies with latitude; apparently geometry varies with altitude.'[21]

Eddington was Britain's foremost astronomer and Director of the Cambridge University Observatory (CUO), where he lived in one wing with his mother and sister. He appeared only sporadically at the High Table, but Milne penetrated his pensive silences and was delighted when Eddington agreed to speak to the Trinity Mathematical Society. Milne was in the chair and described the meeting to his father:

> We held a great meeting of the Trinity Mathematical Society the other night ... Eddington gave a paper on 'The Equilibrium of Gaseous Stars'. ... When the paper was finished and I had made a few rather inadequate presidential remarks, what should I hear but the big booming voice of Rutherford at the back, starting to heckle Eddington. I can tell you I felt my remarks to be less adequate than ever. Eddington and Rutherford, assisted by Hardy, Littlewood, [C.G.] Darwin and Fowler kept up the discussion for one and a half hours, to the no small delight of the mass of undergraduates present.[22]

Milne listened with sharpened attention because he had been approached about a job at the Solar Physics Observatory (SPO) by its founding director, another Fellow of Trinity, Hugh Newall.[23] Sporting a shock of white hair and twenty-five years older than Eddington, Newall was

---

[21] Milne's notebook, in the author's possession.
[22] Letter from Milne to S.A. Milne, 9 November 1919, EAM, MS.Eng.misc.b.423.
[23] Hugh Frank Newall (1857–1944). Astronomer and benefactor. Professor of Astrophysics at Cambridge from 1913 to 1928.

dignified and wealthy; he cut a dash when his coachman drove him through the Cambridge streets in his brougham and pair.

When Hill had heard that Newall was looking for an assistant he recommended Milne. That Milne was totally ignorant of stars and lacked technical know-how about telescopes did not deter Newall, who had fingertip skill for making delicate adjustments to precision instruments. He liked Milne's 'straightforward simple ways'[24] and offered him the post of Assistant Director with the intention that he devote himself to theoretical astronomy. So it was through chance and circumstance that Milne tumbled into astronomy via ballistics and meteorology.

Far from leaping at this offer, Milne did not want to be burdened by a job. He wanted to savour his new-found freedom and even showed signs of rebellion. Until this point, Milne had been on a treadmill. After striving to get to Hymers and Cambridge he had applied himself unremittingly throughout the war. He felt he deserved a pause and pleaded that he wanted to recoup his lost education to which Hill gently responded that one's education gets more incomplete as time goes on. A compromise was reached; Milne would start at the SPO in January in order to learn some astronomy, but without pay, and begin properly on 1 October. He became familiar with the characteristics of well-known stars, but gazing up at constellations in the night sky did not fill him with wondrous awe. For him the beauty of the stars lay in their complicated concoction of mathematical conundrums.

The SPO consisted of a small cluster of domes and a modest brick building close to the CUO — they amalgamated in 1946. The SPO arose from the gift by Newall's father of a large refracting telescope, and expanded with instruments from an observatory in South Kensington, when its site was earmarked for that of today's Science Museum. Newall and Milne had ground-floor rooms on either side of a small library. On the floor above them was the physicist Charles Wilson,[25] small, patient, unassuming, who invented his eponymous cloud chamber for displaying tracks of atomic particles. At the SPO he could pursue his fascination with

---

[24]Letter from H.F. Newall to A.V. Hill, 30 January 1920, AVHL I.
[25]Charles Thomson Rees Wilson (1869–1959). Lecturer, reader and Jacksonian Professor of Natural Philosophy at Cambridge. Nobel Laureate 1927.

atmospheric electrical phenomena, such as thunderstorms, without being troubled by the electromagnetic fields that strayed round the Cavendish.

Wilson was the president of the erudite Cambridge Philosophical Society, founded in the nineteenth century by the geologist Adam Sedgwick[26] and the botanist John Henslow.[27] By the twentieth century the focus had swung from natural history to the physical sciences. Prominent among its Fellows were Rutherford, Newall and the card-playing fraternity of Aston, C.G. Darwin, Fowler and Hardy. Once again Milne was well placed. He was fortunate that they could all vouch for him and in January 1920 he was elected one of their number. This was at a most auspicious moment, just when he needed to get to grips with a new science, for, as a Fellow, he had privileged access to the society's superb library of scientific journals. Certainly the library was the finest of its kind in Cambridge, probably in England, because the society had the foresight to exchange its *Proceedings* for publications from foreign institutions. (Today the collection is the Scientific Periodicals Library, affiliated to the University Library.)

Although Newall was an astronomer of the old school, measuring and classifying stars, he recognised that astronomy was entering a new phase concerned with the internal physics of stars. At this stage astrophysics was in its infancy and possessed none of the glamour that it conjures up today. On the contrary, it was disdained as a hybrid subject, nestling between physics and chemistry and cemented by mathematics applied to stellar data. During the decade, astrophysics matured and became respectable.

Newall was ready to leave theoretical problems to Milne and to let him concentrate on them. Newall excused him from all duties connected with the observatory's telescopes — just once Milne helped Newall rule some diffraction gratings but otherwise never went near the instruments. Aware of Milne's work on the Earth's atmosphere, Newall suggested that Milne might like to investigate stellar atmospheres, namely the outermost layers of stars.

---

[26]Adam Sedgwick (1785–1873). Appointed professor at Cambridge in 1818. Vice-Master of Trinity College.

[27] John Stevens Henslow (1796–1861). Professor at Cambridge from 1822 to 1861.

Milne thrived in the cohesive community of the two observatories. Newall had founded the Observatory Club for the staff to meet over a cup of tea and biscuits — chocolate ones if there was something special to celebrate. Newall left Milne to his own devices, and if he wanted guidance he had merely to step across the grass to consult Eddington and benefit from his uncanny, almost infallible insight. In his cluttered study, Eddington, pale and thoughtful, would listen silently, staring out of the window, before responding laconically. Milne put store by his superb judgement. They enjoyed a close relationship and Eddington entrusted him with checking the calculations and correcting the proofs of his masterpiece *The Internal Constitution of the Stars.*

Astrophysics presented a hotchpotch of interwoven problems. What is the source of a star's energy and how does it move towards the surface of a star to make it shine? How is the life cycle of a star related to its energy? The emerging ideas about atomic structure provided the means to investigate the ionised particles in the atmosphere of a star. Conversely, as Milne pointed out, 'Astrophysics has something to contribute to the physics of the atom.'[28]

He relished the challenge of a fresh field but was handicapped, like all scientists of his day, by a lack of facts. The neutron had not yet been discovered and the 'Bohr atom' consisted of two particles, protons and electrons. The only known forces were gravity and electromagnetism. To set the scientific background more fully, the planet Pluto, subsequently downgraded, had not been located, nor was the expansion of the Universe identified. Radio astronomy, black holes, dark matter, sputniks and spaceships, not to mention astrobiology and extra-solar planets, lay in the future.

Research style differed markedly from that of today. In all but a few papers Milne was the sole author, whereas nowadays teamwork is the norm. There being no computing facilities for simulating mathematical models, Milne simply unscrewed his fountain pen, the one that Richmond had given him, and thrashed out his ideas in longhand.

The stream of fundamental papers on stellar atmospheres that Milne produced in rapid succession is his most enduring work. He summarised them in a 190-page monograph, *Thermodynamics of the Stars,* which

---

[28]Milne, E.A. 'Recent Work in Stellar Physics' in *Proc. Roy. Soc.* 36 (1924), pp. 94–113.

became the indispensable standard reference. A digest of it was translated into Russian. Although the label 'Milne's problem' is sometimes given to the distribution of radiation inside a hot gravitating star, the bulk of his findings do not bear his name because they are embedded within the foundations of astrophysics, a process known as 'incorporation'.

Thermodynamics held the key to understanding the mechanisms that propel energy outwards to the surface of the star, and the relationship connecting the state of ionisation of charged particles in the star's outer layers to their temperature, pressure and density. His starting point was the radiation theory set out by the German astronomer Karl Schwarzschild[29] and the British spectroscopist Arthur Schuster.[30] Milne showed how to modify their approximate solutions of their equations so as to satisfy the requirement that the net flow of radiation energy through the star's atmosphere be constant. The resulting Milne–Eddington integral equation yields a relationship between the temperature of the gas and the absorption of radiation through it. The equation stimulated huge interest well beyond its stellar context. The challenge to find solutions attracted pure mathematicians. It spurred John Littlewood to speak on it to the Cambridge Philosophical Society; it activated papers from G.H. Hardy and his pupil Edward Titchmarsh[31] and prompted research that culminated in the Wiener–Hopf solution.

After predicting the temperature distribution in a stellar atmosphere, Milne revised the law that describes the darkening across the Sun's disc, from the centre to its edge, the limb. His findings were confirmed independently by the Swedish astrophysicist Bertil Lindblad.[32] Milne spotted the implication of reversing the procedure, and pioneered the method of using observations of the darkening to calculate temperatures. In 1922 this work earned him a Smith's Prize — he used £30 of the money to buy a typewriter.

[29]Karl Schwarzschild (1873–1916). Director of Göttingen Observatory, then the Potsdam Observatory.
[30]Sir Arthur Schuster (1851–1934). Professor at Manchester from 1881 to 1907. Foreign Secretary of the Royal Society from 1920 to 1924.
[31]Edward Charles Titchmarsh (1899–1963). In 1931, after appointments at London and Liverpool, he succeeded Hardy in Oxford's Savilian Chair.
[32]Bertil Lindblad (1895–1965) was appointed Director of Stockholm Observatory in 1927.

Milne was no ivory tower mathematician. He needed to discuss his ideas with other people and Fowler was his intellectual whetstone. Milne loved the thrill of attacking problems and solving them with Fowler in a process of cross-fertilisation. Fowler 'tossed back an idea with lightning speed and you had to be agile to field it properly'.[33] They had already disposed of a question left over from their Brigand days, calculating the ideal shape for the orifice of a siren to emit a 'pure' tone. Now, in a major breakthrough for astrophysics, they separated the physical properties of a star's atmosphere from its chemical constituents. Spectroscopy, the analysis of spectra from photographic plates, was in its heyday. Each chemical element has a unique spectroscopic pattern, its 'signature', and stellar spectra reveal the elements that are present. Niels Bohr explained these distinctive patterns by quantifying the parcels of energy required to move an electron from one orbit to another in an ionised atom. Listening to him lecture on quantum theory stimulated Milne 'beyond belief'.[34]

The Indian astrophysicist Megnad Saha[35] had extended Bohr's analysis by elucidating the cause and variation in intensity of the black absorption bands created when white light passes through an ionised gas. Saha's ideas fascinated Milne. Ruminating on them with Fowler, while pacing round his rooms in Trinity, Milne was suddenly seized by a thought. Saha focused on the appearance of the faintest Fraunhofer line in a given series. In a flash of lucidity, Milne realised that the strongest line deserved more attention, and, moreover, was much easier to identify. He had reason to believe that it depended only on the star's surface temperature, unlike the faintest line, which also depended on the abundance of the particular chemical element. In other words, the strongest line was an indicator of temperature, and, if this were the case, would be different for stars of different temperatures. Fowler and Milne put the gist of this to Bohr who encouraged them to explore it in detail. Applying the powerful methods of statistical mechanics, developed by Fowler and C.G. Darwin, everything fell into place. This was a huge step in creating a stellar temperature scale.

---

[33] Milne, E.A. 'Ralph Howard Fowler' in *Obit. Not. of Fellows of The Royal Society* 5 (1943), p. 67.
[34] Letter from Milne to G. Milne, 17 March 1922, in the author's possession.
[35] Megnad Saha (1893–1956). Scientist, politician and social reformer.

Their method of interpreting contour lines in stellar spectra was immediately adopted in observatories worldwide and injected a surge of enthusiasm to spectroscopic studies.

There was more to come. The pressures they calculated were a shock, far, far lower than expected, only about one thousandth of that of the Earth's atmosphere, with the implication that radiation pressure dominated in a stellar atmosphere. Furthermore, in an ingenious reversal of Saha's procedure, Milne and Fowler showed how to assess the chemical abundance of a star — again, with startling results. To the disbelief of the majority of astronomers, enormous quantities of hydrogen seemed to be present, whereas hitherto the composition of stars was assumed to be more or less the same as the Earth's crust.

For a novice in a complicated subject, Milne made his mark astonishingly quickly. His sequence of groundbreaking papers established his reputation, which was always greater internationally than nationally. In part this was due to Newall, who was a great ambassador for astronomy and always made sure that Cambridge was a port of call for astronomers from overseas. When the university accepted his father's telescope it imposed the condition that Newall live nearby, and he built a handsome house, Madingley Rise, with stables and a small farm.[36] The Newalls — his wife was a pianist — made Madingley Rise a centre of gracious hospitality, and Milne noticed that the cosmopolitan conversation encompassed art, music and botany. Milne met many of their foreign visitors, among them three notable Americans: Albert Michelson,[37] who disposed of the non-existent 'ether' in the famous Michelson–Morley experiment; George Ellery Hale,[38] the founding director of California's Mount Wilson Observatory; and Henry Norris Russell,[39] from Princeton, equally accomplished as an observer and as a theoretician.

---

[36] A Dutch astronomer referred to the mahogany-coloured pigs as mapigany hogs.

[37] Albert Abraham Michelson (1852–1931). The first American Nobel Laureate, in 1907.

[38] George Ellery Hale (1868–1938). Inventor of the spectroheliograph. Professor at the University of Chicago from 1893 to 1904. Director of Mount Wilson Observatory from 1904 to 1923. He raised funds for the large telescopes at Chicago, Mount Wilson and Mount Palomar.

[39] Henry Norris Russell (1877–1957). Remembered for the Hertzsprung–Russell diagram, which plots stellar magnitude against temperature and colour. Director of the Princeton Observatory from 1912 to 1947.

Newall was good at bringing on a younger man and saw that Milne wrote reports and sat on committees. Six months after formally starting at the SPO, Milne was elected to the Royal Astronomical Society (RAS) and to its elite dining club the following year. The society evolved from a meeting in a London tavern in 1820, received its royal charter in 1831 and settled into its current premises in Burlington House in 1874. It was the heart and conscience of astronomy in the Western world and its meetings were Milne's lifeblood. He was conspicuously active in its affairs and between the wars there was hardly an issue of *The Observatory* that did not contain a contribution from him, be it a paper, letter or review.

Although he grew fond of Newall and his courtly ways, Milne had his patience sorely tested when Newall, impervious to the need to get papers published promptly, let one of Milne's languish unread on his desk. When Newall got around to approving it and the paper was published, Milne was justifiably irritated that the Dutchman Henrik A. Kramers[40] had anticipated some of his results.

In his evenings, by way of light recreation, Milne mixed with likeminded souls at scientific clubs to which entry was by election only. Again and again, in a merry-go-round of stimulating meetings, he enjoyed the company of a small group of people who became professional friends. In the privacy of a member's room they gave and heard each other's papers, papers that often contained the germ of the big idea by which its author would come to be best known. Besides the Trinity Mathematical Society, Milne was an ardent devotee of the $\nabla^2 V$ Club and the Cambridge University Natural Science Club, and rose to the position of president in both. The $\nabla^2 V$ Club, limited to twenty-five members, was a top-level forum for high-powered mathematical physicists such as Appleton, Eddington and G.I. Taylor. Sometimes Milne used the club as a sounding board, a dress rehearsal for presenting a paper to a professional audience such as the RAS.

Of quite a different character was the recondite Cambridge University Natural Science Club that embraced all sciences, from the study of spiders to quantum theory, from colour vision to aeronautics. It comprised a dozen junior members of the university who were obliged by the rules to

---

[40] Hendrik Anthony Kramers (1894–1952) held appointments at Copenhagen, Utrecht and Leiden.

leave when they attained MA status. This ensured a healthy throughput of new members and created a big pool of past members. A candidate was vetted by his being brought along as a guest in the first instance and Milne was introduced by a young Fellow (later Master) of Trinity, Edgar Adrian,[41] who pioneered the understanding of electrical impulses in the nervous system. The club exacted serious commitment from its members since it met on Saturday evenings and expelled those whose attendance was irregular. The sessions began with whales (sardines) on toast and woe betide the host who failed to provide them! The club's blend of larky humour and friendly criticism appealed to Milne, and hearing members' papers broadened his knowledge, especially in biology. One summer at a punting party on the River Cam, he entertained the club with a talk on heraldry, but his paper titled 'The Relations of Mathematics to Science'[42] provoked accusations of disloyalty. He had implied that mathematics was superior to science and cited Eddington, who claimed that in Heaven scientists could not carry out experiments because they lacked apparatus, whereas mathematicians could still enjoy themselves with abstract numbers. Feigning offence, the club charged a sub-committee to investigate Milne's attitude, and after a piqued exchange of letters, meticulously recorded in the minute book, he was allowed to remain a member.

To celebrate the 1,000[th] meeting in January 1920, members bore the expense of giving a grand dinner for eighty people at St John's. Milne was entranced by the long, narrow, low-ceilinged combination room, exquisitely lit by candles, where the exalted company of past members, including the Master of Trinity, partook of a six-course meal. (For its 100th meeting in May 1922 the $\nabla^2 V$ Club gave a dinner in the same elegant room.) When Milne moved on to the Cambridge Graduate Science Club, he trusted that the Natural Science Club would 'flourish like a pruned tree'.[43] In its centenary year, 1972, the club produced a booklet of past

---

[41] Edgar Douglas Adrian, Baron Adrian (1889–1977). Physiologist. President of the Royal Society from 1950 to 1955. Master of Trinity College from 1951 to 1965. Nobel Laureate 1932.

[42] See Rebsdorf, S. and Kragh, H. 'Edward Arthur Milne: The Relations of Mathematics to Science' in *Studies in History and Philosophy of Modern Physics* 33 (2002), pp. 51–64.

[43] Milne's letter, containing this sentiment, is dated 11 October 1922 and is in the minute book of the Cambridge University Natural Science Club, Scientific Periodicals Library, Cambridge.

members: four held the Order of Merit, ten were Nobel Laureates, and 119 were Fellows of the Royal Society.[44]

Milne seldom encountered mixed company except at private houses. Socially immature, he was at a loss as to how to behave when he muddled the date of a luncheon party given by the mathematician W.W. Rouse Ball,[45] a genial Fellow of Trinity. Rouse Ball was an authority on the history of mathematics and spoke to the Trinity Mathematical Society about Japanese mathematics. He was well-to-do and at his house, Elmside, in Grange Road, he had a squash court and a maze. At his famed luncheons he interspersed his principal guests, heads of colleges, professors and their wives, with up-and-coming younger people. Milne was crestfallen to find he had got the wrong day but Rouse Ball, a perfect host, pressed him to stay and the servants laid an extra place at the table. When Milne left, uttering his thanks as he struggled into his coat, Rouse Ball clapped him on the shoulder, 'Never mind, Milne, see you come again next Tuesday.' The story was fodder for Cambridge gossip, and I remember my father telling it against himself.

Rouse Ball, whose *Mathematical Recreations and Essays* went into thirteen editions, had a penchant for puzzles and gimmicks and liked to show off his latest toy. One such acquisition was Radio Rex, a mechanical dog activated by noise, who came out of his kennel when summoned. Rouse Ball became more than a little testy when Radio Rex kept popping out unbidden. G.I. Taylor spotted that the culprit was J.B.S. Haldane,[46] full of self-importance, who, with his loud rasping voice, was absorbed in conversation. Once he was evicted from the room, Radio Rex behaved himself.

During his post-war Trinity days Milne ceased attending College Chapel. That he turned away from the Christian faith created a painful rift with his father which never had a chance to heal because in April 1921

---

[44]Pepys, M.B. 'Cambridge Natural Science Club 1872 to 1972' in *Nature* 237 (9 June 1972), pp. 317–319.
[45]Walter William Rouse Ball (1850–1925). Mathematician and benefactor.
[46]John Burdon Sanderson Haldane (1892–1964). Socialist, mathematical biologist and author. Lecturer at Cambridge from 1922 to 1933. Professor at University College London from 1933 to 1957.

Sidney died from heart failure. Arthur immediately came to his mother's rescue and took financial responsibility for her and for Philip. Geoffrey was about to sit his chemistry finals at Leeds University — he got a first class — and was not yet earning. Unlike his older brothers, Philip had not won a scholarship to Hymers, but he had started there and was in his first year. Keen that he should continue, my father undertook to meet his school fees and later paid for his university education as well. Fortunately my father was well placed to make this expenditure, for, besides his Trinity fellowship, his SPO salary was £300 and he earned a further £50 a year lecturing in astrophysics. This amounted to a sizeable income for a young man of twenty-five, but, even so, permanently shouldering the expenses of the Hessle household was a substantial commitment. That summer he took his mother and Philip on a motorbike holiday. Having bought Fowler's machine secondhand, on which Philip rode pillion, my father added a sidecar for his mother. I recall the pleasure with which my father expounded on the dynamical differences between a two-wheeled motorbike and a three-wheeled combination.

Generally he took holidays with his Cambridge friends. At Arosa in Switzerland he skied with Aston, Fowler and G.I. Taylor. He went sailing on the Solent with the physicist Patrick Blackett,[47] a man of overpowering personality, who had served in the Royal Navy before taking up physics and becoming an expert on cloud photography and much else. Milne's preference, though, was for walking and climbing. Twice he enjoyed strenuous weeks with the select Trinity Lake Hunt at Seatoller House in the beautiful scenery of Borrowdale. Among his doughty companions were Adrian, C.G. Darwin, Fowler and the Duff brothers, James[48] and Patrick,[49] pillars of the Magpie and Stump. Milne had a day as 'hare', trying to elude the rest of the party in pursuit. The hunt finished with tea at

---

[47]Patrick Maynard Stuart Blackett, Baron Blackett (1897–1974). During World War II he was Chief Advisor to the Admiralty. President of the Royal Society from 1965 to 1970. Nobel Laureate 1948.

[48]Sir James Fitzjames Duff (1898–1970). Lecturer and professor at Manchester from 1927 to 1937. Vice-Chancellor of Durham University. Mayor and Lord Lieutenenant of Durham.

[49]Patrick William Duff (1901–1991). Appointed Regius Professor of Civil Law at Cambridge in 1945.

Gatesgarth or Buttermere, often with everybody soaked to the skin, but much restored after dinner by a sing-song around the piano.

He risked illicit roof-climbing along so called 'routes', clambering up college drain pipes and negotiating parapets, to prepare for a demanding climbing holiday on the Isle of Skye with Aston, an experienced climber, and C.G. Darwin. The fourth man of the party was Lawrence Bragg,[50] who with his father pioneered the use of X-ray diffraction to reveal atomic structure. Had the ropes broken on the gruelling Cuillin, a jagged range of black barren rock, physics would have been the poorer. In every sense Milne was the junior member of the party. The rest were Fellows of the Royal Society, two were Nobel Laureates, and Bragg, who had been in charge of sound-ranging during the war, had won a Military Cross. Milne was in his mid-twenties, Bragg and Darwin in their thirties, and Aston, who supplied confidence when the going got tough, was in his forties. None of them were pleased when 'a rather unattached no longer very young lady'[51] persisted in trailing them. Hoping to give her the slip, they trekked across the moors from Sligachan to the hamlet of Glenbrittle, where they took rooms in the postman's house, only to find she had followed them there. One ruse after another failed to put her off until they devised a winning but uncomfortable strategy. First thing each morning they waded across a deep river, which left them wet through for the rest of the day; they had brought only one change of clothing and kept it dry for the evening.

In 1922 two events immersed Milne more deeply in the world of science. The week-long celebrations in May for the centenary of the RAS thrust him into the company of astronomers from home and abroad. In his presidential speech Eddington noted the trend towards studies of the physics of stars.

In September the British Association for the Advancement of Science (BAAS) met in Hull. Milne must have been proud that his home city,

---

[50] Sir William Lawrence Bragg (1890–1971). Professor at Manchester from 1919 to 1937 and at Cambridge from 1938 to 1954. In 1954 he became Director of the Davy–Faraday Laboratory at the Royal Institution, London. Nobel Laureate 1915.
[51] W.L. Bragg's autobiographical notes, RI MS WLB Box 87, p. 45, courtesy of the Royal Institution of Great Britain.

brightened with flowers, gave the delegates a warm welcome and free transport. Mr Gore took a prominent part in the proceedings and entertained the delegates at a garden party at Hymers. Milne introduced his mother to Hill, who chaired a discussion at which W.S. Tucker explained how he had adapted his hot-wire microphone to test hearing. Milne made his debut as a BAAS speaker, and described his 'cone of escape' for particles ejected from the surface of a star: the higher the temperature and the weaker the gravitational field, the greater the speed of the particles. Religion had its place in the programme too. Dean Inge compared current knowledge about heavenly bodies with what the Psalmists had to say and the Archbishop of York preached on the comradeship of science and faith.

In the late spring of 1924, when his star was in the ascendant, Milne's glorious existence abruptly ended. While riding his motorbike he realised that he was seeing double and he had to cover one eye with his hand to finish his journey. Adrian, who had advised him about reading glasses, correctly suspected that he was a victim of the pandemic of *encephalitis lethargica*, or 'sleepy sickness',[52] which often attacked young people. Troubled vision was a common symptom. The life-threatening disease, which swept westwards across Europe to America and then mysteriously vanished, caused inflammation and destruction of the mid-brain and basal ganglia. After a flu-like fever the patient sank into a deep coma, some never to be roused, and the protracted convalescence was marked by relapses. One in three patients died. Others were left speechless, motionless or with personality disorders, and the government built institutions for children who were permanently affected.

Adrian arranged for Milne to be seen by the leading authority on the disease, the distinguished neurologist George Riddoch,[53] a humorous Aberdonian. Luckily Milne was only mildly afflicted and he had excellent care, thanks to his Brigand friends. With unobtrusive kindness, William Hartree and his indomitable wife Eva[54] took him into their house, Lyghe, in Newton Road in Cambridge. It was typical of Eva Hartree, small,

---

[52]Not to be confused with the disease transmitted by the tsetse fly. Oliver Sacks describes patients suffering from *encephalitis lethargica* in *Awakenings*, London: Picador, 1982.
[53]George Riddoch (1888–1927). Neurologist.
[54]Eva Hartree (1873–1847). Social reformer. President of the National Council of Women.

bright-eyed, with formidable commitments in public life and soon to be the city's first lady mayor, that without ado she absorbed a sick man into her busy household. Vitally for Milne, the Hartrees shook him awake to prevent his drowsing off into a heavy stupor. When he was on the mend, he tottered round to the younger Hartrees, Douglas and Elaine, in Bentley Road, where their daughter would sit on Milne's knee, fiddle with his watch chain, and address him as 'Grandfather'!

Richmond took him for little excursions in his car, and when Milne was well enough for the long drive to Hessle, he drove him home. (Sadly, gunnery trials had cost Richmond his hearing and he could only manage one-to-one conversations.) Foolishly they detoured to see as many cathedrals as possible, starting with Peterborough, so that two days later when they reached Hessle Milne was utterly exhausted and had to retire to bed.

A major illness can be a life-changing event, and those who knew Milne before and afterwards detected a slight change. He lost something of his effervescent bouncy energy, and his demeanour became more stubborn, more disputatious. Mercifully the disease did not impair his capacity for creative thought, nor did he lose his copious flow of rapid speech, at least not for the present.

While recovering, Milne was offered the Beyer Chair of Mathematics at Manchester University. This was an obvious promotion and, at the age of twenty-eight, a considerable compliment, but he was reluctant to abandon his plans for the coming year. He had resigned from the SPO in order to be mathematics lecturer at Trinity in succession to G.I. Taylor, a position Milne greatly esteemed as he felt it would endorse his legitimacy as a mathematician. He would continue to lecture in thermodynamics and in astrophysics, which he had constantly to revise to keep pace with rapid developments. He brooded fretfully over whether to accept the Manchester chair and after tortuous indecision did so with the proviso that he postpone starting for a year, until autumn 1925. Not only could he fulfil his Cambridge wishes but he would regain his health more thoroughly, the better to cope with the demands of a new job.

In the golden age of physics, theoretical astronomy was coming into its own and Milne was at its cutting edge. His happy and productive Cambridge years came to a climax in July 1925, when the International Astronomical Union met in Cambridge. At the inaugural session in Rome in 1922 there were ninety astronomers from fifteen countries. This time there were twice as many delegates from some twenty countries. They represented a substantial proportion of all the world's astronomers and they tended to know one another personally. Milne found especial rapport with the Dutch and the Americans. Besides papers on astronomy, the nine-day event included an address by the Chancellor of the university, Lord Balfour; a service in King's College Chapel; and a garden party hosted by the Newalls at Madingley Rise.

On his last night in Cambridge, Milne went to bid farewell to Newall at the SPO. The cloudless summer evening was perfect for looking at stars and one of the staff asked, 'Well Milne, before you leave us, would you like to look through a telescope?', which was greeted with a gasp of astonishment that he had never before done this.

The next week, standing beside Mr Gore at the Hymers speech-day, Milne presented the prizes. Swotty Blinks had mutated into the guest of honour.

Chapter 6

Riding on a Sunbeam

The important thing in science is not so much to obtain new facts as to discover new ways of thinking about them.[1]

The rhythm of Milne's life at Manchester was radically different from Cambridge, academically, socially and domestically. Away from the cosseting of college servants in Trinity's sequestered masculine courts, he had to fend for himself. He found lodgings at 7 Clyde Road in West Didsbury, an area popular with university staff, only a short tram ride from the university in Oxford Road. His landlady cooked his evening meal from such food as he had purchased and he ate lunch in the informal and friendly atmosphere of the university refectory, where seniority did not dictate where people sat. It was easy for a newcomer to settle in and as Rutherford put it, 'At Cambridge one might sit next to a man for forty years without knowing whether he had a wife, mother or sister, whereas before he had been at Manchester a week he knew his colleagues and their wives and was calling their children by their Christian names.'[2]

Among the staff of less than two hundred men and women, Milne already knew a handful who had migrated to Manchester from Cambridge. Foremost among them was Lawrence Bragg, Dean of the Faculty of Science, who established a pre-eminent school of crystallography. He and his wife, Alice, a cousin of Bertram Hopkinson, did Milne a great kindness in caring for him after a surgical operation. No sooner had he thrown off the last lingering vestiges of his sleepy sickness than he suffered a painful double mastoid of the ear. Without modern antibiotics, the treatment was to drill the bone to drain the abscesses. The Braggs collected

---

[1] W. L. Bragg quoted in Mackay, A.L. (ed.). *Scientific Quotations*, Bristol: Adam Hilger, 1991, p. 38.

[2] Eve, A.S. *Rutherford*, Cambridge: Cambridge University Press, 1939, p. 350.

him from hospital and saved him from a lonely convalescence by taking him to their home. Alice Bragg changed his dressings daily.

Someone with whom Milne spent considerable time, a bachelor like himself, was the Cambridge physicist Reginald James,[3] whom Bragg had recruited to his laboratory. Before the war James had survived an ordeal of unimaginably severe hardship. On a whim, after two lacklustre post-graduate years at the Cavendish, he had joined Ernest Shackleton's ill-fated expedition to the Antarctic. Pack ice crushed their ship the *Endurance* leaving the party stranded on Elephant Island, where there was no certainty that they would ever be rescued. During the dreadful months of waiting, surviving on penguin meat, James's sense of humour bolstered morale.

If Milne wanted congenial company and relaxation he could drop into the elegant premises of the Literary and Philosophical Society at 36 George Street (destroyed in the Blitz) for an afternoon cup of tea or a browse in its library. The society served as a social centre for its members, a mixture of academics and industrialists, and was a bastion of lively dis-cussion across a broad range of subjects, religion and politics excepted. Milne sat on its council, was secretary during Bragg's presidency, and gave more than one talk.

The 'Lit and Phil' serves as an example of Manchester's strong harmo-nious links between education and commerce. Unlike Cambridge, there was no snobbish discrimination between the university and the city. Indeed, long before the university took shape, the city had valued science, and laid claim to the eminent chemist John Dalton[4] and his pupil James Joule.[5] That the grimy city relied on trade for its wealth in no way dimin-ished its enthusiasm for cultural pursuits nor compromised its liberal independence, evidenced in the *Manchester Guardian*.

---

[3]Reginald William James (1891–1964). He became a professor at Cape Town University in 1937. President of the Royal Society of South Africa from 1950 to 1953.

[4]John Dalton (1766–1844), chemist and meteorologist, enunciated the principles of chemi-cal combination and formulated gas laws. President of the Manchester Philosophical Society from 1817 to 1844.

[5]James Prescott Joule (1818–1889), experimental physicist, established the principle of the conservation of energy and gives his name to the SI unit of energy.

The university grew from an earlier institution, Owens College, founded against the backdrop of nineteenth century non-conformism. John Owens,[6] a somewhat crotchety bachelor, opened up tertiary education to Mancunians who were excluded from the current universities by reason of religion, day-time employment, or the high cost of living away from home. The new university was not constrained by archaic traditions and put students first. It provided part-time as well as full-time courses, during the day or in the evening to suit its students' needs.

By the twentieth century the unpretentious university was in full flower. Its formidable standing in the humanities derived from the lawyer-politician Sir Alfred Hopkinson,[7] and three Balliol men, historians Thomas Tout[8] and James Tait,[9] and the philosopher Samuel Alexander.[10] (Milne attended the unveiling of Jacob Epstein's[11] bust of Alexander, which fronts the Arts Building.) In the physical sciences Manchester was second only to Cambridge. Its most eminent practitioners were the chemists Sir Henry Roscoe,[12] William Perkin[13] and Arthur Lapworth,[14] and the physicists Niels Bohr, Ernest Rutherford, Hans Geiger, Ernest Marsden[15] and latterly Lawrence Bragg. Sir Arthur Schuster, who retired early to make way for Rutherford, had promoted astronomy, but enthusiasm had waned and Milne did not have a single astronomical colleague. Without ready access

---

[6]John Owens (1790–1846). While working in his father's millinery business he made shrewd investments in the railways that enabled him to leave nearly £100,000 to found a college free of religious tests.

[7]Sir Alfred Hopkinson (1851–1939). Vice-Chancellor of Manchester University from 1900 to 1913. Member of Parliament from 1895 to 1898 and from 1926 to 1929.

[8]Thomas Frederick Tout (1855–1929). Professor at Manchester from 1890 to 1925.

[9]James Tait (1863–1944). Lecturer and professor at Manchester from 1887 to 1919.

[10]Samuel Alexander (1859–1938). Professor at Manchester from 1893 to 1924.

[11]Sir Jacob Epstein (1888–1959). Sculptor. Born in the USA, he became a British citizen in 1911.

[12]Sir Henry Enfield Roscoe (1833–1915). Professor at Owens College from 1857 to 1885.

[13]Sir William Henry Perkin (1860–1929). Appointed to a chair at Manchester in 1892, and at Oxford in 1912.

[14]Arthur Lapworth (1872–1941). Lecturer and professor at Manchester.

[15]Hans Wilhelm Geiger (1882–1945). German physicist. While at Manchester with Rutherford, Geiger and Sir Ernest Marsden (1882–1945), a New Zealander, invented a device for counting alpha particles, namely the Geiger counter.

to data, to which he was accustomed at Cambridge, he asked the great American observatories, Lick, Harvard and Mount Wilson, to put him on their mailing lists to receive their publications.

The high reputation of the Mathematics Department sprang from Horace Lamb,[16] whose widely read textbooks exerted a profound influence on the teaching of mathematics. Manchester gave Milne a taste of working in a departmental centre, and he liked the way it engendered identity and cohesion. The department had two professors (today, the enlarged university has nearly thirty), with Milne in charge of applied mathematics and Louis Mordell[17] of pure. Mordell was a keen walker, and organised exhausting all-day departmental rambles in the Derbyshire hills, which began with a train ride to some remote spot in the countryside. There would be a fair distance to be covered on foot before reaching the station for their return, where they would tuck into high tea before collapsing with relief into the homeward train.

The professors, supported by a staff of six, taught some three to four hundred students, of whom about twenty might be honours finalists. Lectures were followed by questions marked by the lecturer. The professors were spared this task, however. Doris Withington, an assistant lecturer who doubled as the department's secretary, did their marking. She stands behind Milne in Fig. 9.

Milne soon established himself with the students, and impressed Mordell with his energy and mental alertness. Mordell noticed that besides possessing a photographic memory, whereby Milne had perfect recall of whole pages, he could recite speeches he had heard, word for word.

Although the university accommodated students in halls of residence, many travelled in daily, some from outlying farms and villages up to thirty miles away. They were usually hard pressed, for many were expected to carry out chores at home before settling down to their studies each night. They had scant time for recreation, not even the tame parties of dancing

---

[16]Sir Horace Lamb (1849–1934). Professor at Manchester from 1885 to 1920. Father of the portrait painter Henry Lamb.

[17]Louis Joel Mordell (1888–1972). Fielden Professor at Manchester University from 1922 to 1945. Sadleirian Professor at Cambridge from 1945 to 1953.

and soft drinks run by the Mathematics Department. At these gatherings, according to one student, Milne seemed slightly distrait, musing on whatever was preying on his mind until he jotted down his thoughts on a scrap of paper.

One commuting student shone with outstanding brilliance. The pure mathematician Harold Davenport,[18] shy and cautious, reckoned that his university training scarcely differed from his sixth form days at Accrington Grammar School. At the age of nineteen, having graduated with first class honours, he wanted to continue his studies at Cambridge provided he won a scholarship. Milne encouraged him to sit for Trinity rather than Mordell's college which was St John's, and Davenport's success pleased Milne enormously. It brought to mind his own triumph in similar circumstances. Both came from modest homes, were the first in the family to go to university, and their Trinity scholarships were the gateway to their mathematical careers. When Davenport consulted Milne on the lectures he should attend, Milne warned, 'I don't think it is the job of a young mathematician to devote himself to foundations ... Work on foundations is rarely permanent; paradoxically [another] thinker comes along and says it is all wrong.'[19]

Milne's predecessor in the Beyer Chair was his former supervisor Sydney Chapman who was an ardent champion of using vectors to solve problems. (Vectors specify direction as well as magnitude.) Generally the treatment of vectors in university courses was a mere exercise in manipulating their distinctive notation, often in a slapdash manner, whereas Chapman regarded them as powerful and elegant tools. Milne took up his crusade. He argued that plunging into vectors directly, without dabbling first in Cartesian co-ordinates, was economical and efficient and that the vector solution offered insight into the dynamics of the situation. His favourite example was a spinning top: 'the problem is formulated vectorially, solved vectorially and the vectorial solution proclaims the resulting motion'.[20]

---

[18]Harold Davenport (1907–1969). He held chairs at Bangor, London and Cambridge.

[19]Letter from Milne to H. Davenport, 12 October 1927, papers of H. Davenport G206, Trinity College Library, Cambridge.

[20]For Milne's defence of vectors, see letters from Milne under the heading 'The Hamiltonian Revival' in *The Mathematical Gazette* 25 (1941), pp. 106–108 and 298–299.

When the Cambridge mathematician Bertha Swirles (Lady Jeffreys)[21] joined the department, Milne handed her his lecture notes on vectors. She was puzzled by the number of examples relating to ballistics since he made no mention of his Brigand work. Like Mordell, she was conscious of his intensity, and recalled his talking at a great rate while sitting cross-legged on a desk. She did not dispute his claim that he could think vectorially, and dubbed him a 'whizz-kid' for his knack of selecting the most convenient vectors, eliminating the unwanted ones. Yet Milne's enthusiasm for vectors did not always go down well with his students. In his eagerness Milne raced across the blackboard, his handwriting, never clear at the best of times, deteriorating line by line. If the suffixes became illegible, he forfeited the students' patience.

Milne was determined that his teaching commitments should not diminish his research. To keep up with developments in quantum physics — he always felt he had a poor grasp of the subject — he frequently went to Cambridge where he remained an integral part of its scientific scene.[22] Besides letters and reviews, he published four or five papers a year, a prodigious output that exceeded the combined papers of the rest of the Mathematics Department.

Within a week of his thirtieth birthday, in February 1926, he was thrilled to attain the highest echelon to which a British scientist can aspire. Sponsored by H.F. Newall, A.S. Eddington, F.W. Aston, G.I. Taylor, C.T.R Wilson and others, he was elected a Fellow of the Royal Society, Britain's most distinguished scientific institution. (The privilege cost him an admission fee of £10 and an annual subscription of £5.) Milne was one of the youngest Fellows to be elected, the average age for mathematicians being forty-two.[23] At the society's rooms, then on the east side of Burlington House, he mixed with the country's top scientists. He served on its council and specialist committees — some of them were responsible for advising

---

[21]Bertha Swirles Jeffreys (1903–1999). She held appointments at Manchester, Bristol, Imperial College London and Girton College, Cambridge, where she was Vice-Principal. Her husband was the geophysicist Sir Harold Jeffreys.

[22]McCrea, W.H. 'Cambridge Physics 1925–1929' in *Interdisciplinary Science Review* 11, no. 3 (1986), p. 271.

[23]Lyons, H. *The Royal Society*, Cambridge: Cambridge University Press, 1944, p. 304.

the government and administering grants, there being no ministry of science, and civil servants being mainly products of literary culture.

Consequent on his analysis of stellar atmospheres Milne turned his attention to the Sun's magnetic storms and aurorae, spectacular streamers of coloured light at the poles caused by the solar wind. Suspecting that they had features in common with nova outbursts, the sudden blazing of a star with huge brightness, he devised a process to account for these cataclysmic disturbances. It made an atom 'go with a wallop,'[24] he told Rutherford, as the atom was expelled from its star at a speed of 1,600 kilometres per second. Milne intended to read his nova paper to the RAS meeting in May 1926. England was in the grip of the general strike, which all but paralysed the country, and the only train running from Manchester to London stopped at every station and took eight hours. But he was not to be defeated and somehow got to the meeting where his paper was warmly hailed as the first consistent explanation of novae. Eddington called it 'a very beautiful paper'.[25]

In 1927 a total eclipse of the Sun caught the imagination of the public and aroused interest in astronomy. It was two centuries since Britain had last witnessed a total eclipse and people were agog about an event of a lifetime. On 29 June the band of darkness, thirty miles wide, would sweep from Criccieth in North Wales across to Hartlepool in County Durham, up the Scandinavian Peninsula to the Arctic Circle, then across Siberia to the North Pacific Ocean. The eclipse brought Milne many invitations to speak and not just to specialist audiences. Besides addressing Manchester's Astronomical Society and Physical Society as well as Hull's Astronomical Society, chaired by the Head of Mathematics at Hymers, Henry Forder,[26] he spoke to teachers in six towns at the request of the Lancashire County Council. At a lecture that Milne gave as part of an extra-mural series in Manchester, every seat in the Chemistry Theatre, the second largest room

---

[24] Letter from Milne to Rutherford, 7 May 1926, papers of Ernest Rutherford, MS.Add.7653, Cambridge University Library.

[25] *The Observatory* 49 (1926), pp. 182–185.

[26] Henry George Forder (1889–1981). In 1934 he left Hymers to become a professor at the University of Auckland, New Zealand. Author of 'Kinematic Relativity and Textile Geometry' in *Quart. J. Math.* 11 (1940), pp. 124–128.

in the university, was taken, and people had to be turned away. Such was the level of excited curiosity.

Milne explained why an eclipse provides an exceptional opportunity to obtain information about the Sun.[27] Usually the Sun looks like a smooth round ball owing to the light coming from its spherical shell, the photosphere. We do not see the two outermost layers: the glowing red chromosphere and the extremely hot corona whose fiery prominences constantly change, giving it an irregular shape. During an eclipse the Moon obscures the photosphere, thereby letting us see these outer layers as a bright halo surrounding a black disc. Just before and just after totality, an emission spectrum of the chromosphere and corona can be obtained, and Milne hoped that this spectroscopic evidence would confirm certain predictions he had made.

The chromosphere intrigued Milne for the clever way it stays wrapped around the photosphere. After much deliberation he concluded that the equilibrium must result from an internal balancing process, since its particles are neither ejected outwards by radiation impulses, nor fall inwards under the weight of their solar gravity, but are suspended between these opposing forces. What particle could meet these criteria?

Ionised calcium, which has a strong spectral line, seemed to be a conspicuous constituent of the Sun's chromosphere and Milne seized on it as the best candidate for this delicate balancing act. The merit of the ionised calcium atom, Milne argued, was the unique behaviour of its spare electron, which exists in two forms and enables it to 'float on a Sunbeam' in Eddington's felicitous phrase.[28] On absorbing an impulse of energy, the spare electron rises to a higher electron orbit where it revolves before spontaneously sinking back to its former orbit. At the next excitation the cycle repeats itself, thereby maintaining equilibrium. Milne calculated that the cycle occurs 20,000 times a second, which agreed with laboratory data. From this Milne calculated the mass of the chromosphere to be about 300 million tons. His ingenious mechanism, original, neat and picturesque, had an appealing simplicity, and was widely accepted because

---

[27]Three years later in the Pyrenees the French astronomer Bernard Ferdinand Lyot (1897–1952) observed the corona in broad daylight with his three-lens coronagraph.

[28]Eddington, A.S. *Stars and Atoms*, Oxford: Oxford University Press, 1927, p. 71.

it seemed to provide a satisfactory explanation of the observational facts. But it turned out to be wrong.

The story of Milne's calcium chromosphere is a good example of his influence on astrophysics. He invented a highly plausible theory that sparked enormous interest. Eddington liked it and sang its praises. It activated correspondence in *The Observatory* and *Nature* from, among others, the Russian astronomer Boris Gerasimovich,[29] later executed in Stalin's purges, and Giorgio Abetti,[30] Director of the Arcetri Observatory near Florence, once home to Galileo. Testing the credence of the theory spawned an avalanche of corrections, embellishments and improvements — until ultimately it was discarded.

When Milne made his proposal, the cause of a strong spectral line was not understood and he mistook it for abundance. A few years later it became clear that the chromosphere is chiefly ionised hydrogen, and that despite ionised calcium producing stronger spectral lines, its contribution to mechanical support of the chromosphere is negligible.

As eclipse fever mounted, the railway companies GWR, LNER and LMS ran special overnight trains so that when dawn broke passengers would be in the band of totality. Newspapers laid on trips and *The Times* chartered a plane from Imperial Airways. Astronomers organised their own expeditions. In the opinion of F.J.M. Stratton[31] DSO, stocky, energetic and veteran of many an eclipse expedition, preparing for one was like preparing for battle.

Milne was not much good at practical matters but eagerly accepted Newall's invitation to join his team in Norway and afterwards to accompany the Newells on a private holiday. Norway offered better viewing of the eclipse than England for two reasons: the duration of totality was longer and the Sun was at a higher elevation. Armed with the Newalls' gift of a Norwegian dictionary, Milne reached Aal, on the main Bergen–Oslo

---

[29]Boris Petrovich Gerasimovich (1889–1937). Director of the Pulkovo Observatory.

[30]Giorgio Abetti (1882–1982). Author of *The Sun*, London: Faber and Faber, 1957 and *The History of Astronomy*, London: Sidgwick and Jackson, 1954.

[31]Frederick John Marrian Stratton (1881–1960). Astronomer, soldier and administrator. In 1928 he succeeded Newall as Director of the Solar Physics Observatory and Professor of Astrophysics.

railway line, a couple of days before the eclipse, where Newall had booked the grounds of a hotel in which to put up his apparatus. Having missed the worst of the arduous preparations — unpacking twenty-five cases of equipment and nine bales of tenting, as well as setting up the instruments — Milne was in time for the final rehearsal of executing photographic procedures with split-second accuracy.

On the morning of the eclipse the party awoke at 4 am to tantalisingly overcast skies. In vain they waited for the weather to improve but to everyone's dismay thick clouds hid the Sun during the vital period of totality. In the glum, frustrated atmosphere of disappointment Milne helped to dismantle the site, but ahead of him was a treat, a luxurious holiday in the Norwegian countryside with the Newalls in their chauffeur-driven car. Milne willingly fetched and carried at their behest, each day loading and unloading the car with a mass of luggage and paraphernalia, along with cherished lenses and prisms, too precious for the consignment going back to England. The grand leisurely tour was a glorious idyll, unsurpassed in Milne's experience, then or later. Newall was a knowledgeable naturalist and whenever his eye lit on a rare wildflower they stopped to identify it, probably kindling Milne's interest in botany.

Once back in England, Milne was gratified to learn that the team from the Royal Greenwich Observatory, who had gone to Yorkshire for the eclipse, had been luckier with the weather. Their results at Giggleswick School, Settle, showed good agreement with his predictions about the change in temperature across the Sun's limb.

The eclipse widened Milne's experiences and brought him into contact with the general public. The county council engaged him again and asked him to speak at the cotton towns of Bury and Colne, north of Manchester. Milne illustrated these talks with slides, which he found bothersome to prepare. It is 'a great sweat', he told A.V. Hill, 'but they seem to like it'.[32]

---

[32]Letter from Milne to A.V. Hill, 24 November 1927, AVHL I.

Chapter 7
New Horizons

Marriage has many pains, but celibacy has no pleasures.[1]

Milne had seldom encountered girls. He came from a family of brothers, attended a boys' school, and thereafter lived in the company of men at Cambridge and during the war. He had wistfully confided to his Trinity friend Tressilian Nicholas,[2] later the Senior Bursar of the college who astutely invested in Felixstowe Docks, that he wished to be as happily married as Nicholas was. Yet, as Nicholas pointed out, 'those with supreme gifts for mathematics seldom used to meet members of the other sex. Neither Hardy, nor for that matter Isaac Newton, ever married.'[3]

Milne knew that Sydney Chapman had found a wife in Manchester and wondered if perhaps he could too. He was shy but there were ample opportunities to meet the opposite sex, for at Manchester, unlike Cambridge, he had women colleagues — and they enjoyed equal status with men. He started taking girls out and eventually had the good sense to replace his noisy, greasy motorbike with a Morris Oxford Saloon, more conducive to his amatory quest. Outings in it helped him to win the affections of Margot Scott Campbell, aged twenty-six, a science graduate of Newnham College, Cambridge, who held two part-time jobs. Intrigued by the comparatively recent discovery of vitamins (originally called vital amines) and by the influence of diet on health, she had recently gained a diploma in institutional housekeeping, and combined lecturing on dietetics at Pendlebury Hospital with teaching sixth form chemistry at Withington Girls' School.

---

[1] Samuel Johnson, *Rasselas*, London: Dent, 1926, Chapter 26.
[2] Tressilian Charles Nicholas (1888–1989). Geologist. Senior Bursar of Trinity College from 1929 to 1956.
[3] Letter from T. Nicholas to the author, 20 February 1975.

My mother was born near Aberdeen, the daughter of a canny Scots advocate, Hugh Fraser Campbell and his wife Jessie (née Fiddes). When Margot was ten her mother died from a botched hysterectomy operation and the family decided that it would be best for her and her elder sister Jean to live with their Fiddes relatives in Manchester. Hugh Campbell was delightful company, clever, with an impish sense of humour, but utterly shiftless and disorganised. The girls returned to him for summer holidays but they greatly preferred being in their aunt and uncle's lively household at 173 Wilmslow Road, where there was an endless procession of visitors.

Jessie's sister Agnes, known as 'Grundy', stout and affable, abounded with humanity and her servants adored her. She was an attentive surrogate mother, quickly integrating the girls into local social activities such as tennis afternoons, and taking them to Gilbert and Sullivan operettas. The girls' favourite treat was Professor Lang's[4] tea club for children. Tall, with a fluffy white beard, he made a fine figure presiding over a table loaded with mouth-watering food. Should a hapless parent appear, he or she was restricted to the bread and butter, while the children laid into the cakes and éclairs called 'Othellos' or 'Desdemonas' according to colour.

Agnes kept house for her brother Edward,[5] a former Pro-Vice-Chancellor of Manchester University, who was its undisputed elder statesman. Spare, angular, laconic, he had won scholarships in classics to gain first class degrees at Aberdeen University and at Peterhouse, Cambridge. During his first post at Owens College, lecturing in Roman history, his acumen and flair for administration became apparent. He was appointed registrar and guided the university to independence. On his retirement the university created for him the Ward Chair of History.

For a while his private secretary was Caroline Lejeune,[6] the future well-known film critic of *The Observer*. She and Jean Campbell were friends

---

[4] William Henry Lang (1874–1960). Palaeobotanist.

[5] Edward Fiddes (1864–1942). Author of *Chapters in the History of Owens College and of Manchester University 1851–1914* (1937). He edited W.T. Arnold's *Studies in Roman Imperialism* (published posthumously in 1906). Arnold was the brother of Mrs Humphry Ward; their uncle was the poet Matthew Arnold.

[6] Caroline Alice Lejeune (1897–1973). Journalist and critic.

and they collaborated on an unpublished novel, *Equal Thanks*, writing alternate chapters to surprise each other over the plot. Caroline admired Edward's irreproachable discretion in diffusing awkward situations. After listening patiently to an outburst from an irate member of the university, he would 'pass a diplomatic hand over his small Macmillan moustache and achieve a masterpiece of saying nothing'.[7]

In their uncle, the girls had an ardent champion of education. With a Scot's belief in its importance, he did all he could for women students at the university and succeeded C.P. Scott,[8] of the famously liberal *Manchester Guardian*, as chairman of the Council of Withington Girls' School, which the girls attended. At first for a year, Margot went to Ladybarn House, a progressive co-educational preparatory school with advanced views. The children monitored a rain gauge, did a lot of craft-work and earned cash rewards for prowess at gym. The tariff for climbing the rope was one penny. Initially she was at a disadvantage because she knew only Scottish history, a passion of her father's, but Edward Fiddes soon put this right. She followed Jean to Withington Girls' School and was an accomplished all-rounder; she acted in plays, captained the hockey team and became head girl. Encouraged by her uncle and showing an aptitude for botany and chemistry, she successfully sat the competitive entrance examination to Newnham College, Cambridge, where Jean had read classics. Their hall tutor was Miss Joan Pernel Strachey[9] (later principal of the college), skinny like her brother Lytton. Of the two sisters Jean was more solemn, more scholarly, and sufficiently ahead of her age group that before going up to Cambridge she completed a year at Manchester University. Margot was jollier, even frivolous. When the girls were given money to buy books she splurged hers on a new hat.

Despite women at Cambridge sitting the same examinations as men, the university granted them only 'titular' degrees and denied them a degree ceremony. In 1921, the fiftieth anniversary of the founding of Newnham, this injustice was again aired and the contentious issue debated

---

[7] Lejeune, C.A. *Thank You For Having Me*, London: Hutchinson, 1964, p. 60.
[8] Charles Prestwich Scott (1846–1932). Owner and editor of the *Manchester Guardian*.
[9] Joan Pernel Strachey (1876–1951). French scholar. Principal of Newnham College from 1923 to 1941.

at the Senate House. The proposal for parity was lost, and afterwards, in an explosion of macho boorish bravado, an unruly rabble of male undergraduates stormed Newnham. The principal, Miss Blanche Athena Clough,[10] daughter of the poet Arthur Hugh Clough and cousin of Florence Nightingale, was a woman of few words. Anticipating trouble she calmly advised her students to get back to college early and Margot was safely inside the locked gates before the wild mob rammed them with a porter's coal trolley. This commotion cut a sorry contrast to the enlightened attitude towards women which prevailed at Manchester and to which Margot was accustomed.

The Fiddeses liked nothing better than entertaining guests to lunch or dinner and their house was the social hub of the university. They saw it as their duty, and a thoroughly enjoyable one, to nurture the university and its staff. In this respect they thought the Rutherfords, who also lived in Wilmslow Road, were wanting. On Sundays the Fiddeses held open-house tea parties and my parents met as Margot handed round the sandwiches and cakes. The romance blossomed and when the Mathematics Department got wind of it, they hatched a mischievous scheme to keep the couple apart during Mordell's next expedition. Throughout the day Doris Withington stuck closely to Milne while the men surrounded Margot, a ruse which no doubt added to their ardour. In November 1927 news of their engagement leaked out at a reception given by the Vice-Chancellor before the formal announcement. The Fiddeses gave an engagement party for which Margot wore a new blue dress and Milne was radiant with joy. The wedding was set for June.

Milne held the Fiddeses in the warmest affection and felt more in tune with them than his own relatives. Some people found Edward Fiddes aloof and severe. He was renowned for salty comment but Milne admired his erudition, unostentatious probity and moral rectitude. They had the same cast of mind, a shared love of literature, history and high ideals. Marriage to Edward's niece put Milne under a spotlight and catapulted him to the heart of the university as if he had married the boss's daughter. The marriage confirmed his progression up the social scale and by now he was entirely at ease in the gracious homes of the Newalls, the Hartrees and the Braggs.

---

[10]Blanche Athena Clough (1861–1960). Principal of Newnham College from 1920 to 1923. She was the niece of the first principal, Anne Jemima Clough.

Thanks to her Fiddes upbringing, Margot was alive to the machinery and pitfalls of university politics and well suited to be a professor's wife. The couple had their Cambridge education in common, and, like Milne, she had participated in the settlement movement by helping to run Newnham's programme for disadvantaged women. But their families had different economic perspectives. She was unacquainted with the stretched budgets of the Milnes and Cockcrofts and used to live-in servants. Moreover, she had no financial worries about meeting her university expenses because she and Jean had inherited small legacies from their Fiddes grandfather, who had risen from junior clerk to joint manager at the North of Scotland Bank. The Fiddeses ate dinner, whereas the Milnes sat down to high tea. Margot had enjoyed foreign travel to Egypt and elsewhere. After Cambridge she had spent some months in Lausanne, popular with young English people for the favourable exchange rate of the pound against the Swiss franc.

Unlike Milne, several of her forebears went to university. Her father distinguished himself at Aberdeen University by coming top of his year in every subject before proceeding to its new bachelor of law degree. His sister Isabella, a woman of marked intellect, taught German and French, and spoke Gaelic. Edward Fiddes's brother, Alexander, a graduate of Aberdeen University, became a minister in the Church of Scotland. Margot's cousins, sons of Isabella and Alexander, were Oxford graduates.

With the prospect of supporting a wife and children, Milne felt it incumbent to take out life assurance. Hitherto, owing to his sleepy sickness, he had been refused, and was therefore immensely relieved to be rated an acceptable medical risk. With every intention of settling in Manchester, he and Margot found an attractive double-fronted house close to the Fiddeses, which he purchased outright. The pleasing lines of 25 Belfield Road graced the dustcover of a book about Edwardian domestic architecture.[11]

Although Milne was happily ensconced at Manchester, he and former Cambridge colleagues, who had gone to appointments elsewhere and missed its stimulating science clubs, were keen not to lose touch and wanted to get together on a regular basis. In February he arranged to meet

---

[11] Long, H.C. *The Edwardian House*, Manchester: Manchester University Press, 1993.

Patrick Blackett, who had supplied him with data for the calcium chromo-sphere, at his mother's house[12] in Hampstead, north London. Along with Douglas Hartree and two others they founded the London Physics Club,[13] which met on four or five Saturdays a year, usually in the rooms of the Royal Society, to exchange ideas and keep up to date with advances in atomic physics. The club quickly garnered members — Lawrence Bragg was one of its early chairmen — and they knew each other sufficiently well to speak freely without risking ridicule if they exposed their ignorance.

In March, at the invitation of Lawrence Bragg's father Sir William Bragg,[14] the Director of London's Royal Institution, Milne gave one of its fashionable Friday Evening Discourses. This was a great compliment as the occasions had considerable social caché. In the raked auditorium the audience, attired in full evening dress, were hungry for the latest scientific news. Yet although they were critical and refined, it was claimed that they possessed knowledge and ignorance in equal parts. Milne explained the structure of the outermost layers of the Sun, and ended, 'Somehow or other the Sun knows how to arrange its layers … and in so doing pro-pounds an attractive puzzle for the mathematician and solar physicist.'[15]

At the end of May, Milne set off on a whistle-stop lecture tour in Holland, to Leiden, Amsterdam, Utrecht and Gröningen in the far north. On his return on 5 June, three weeks before the wedding, he found a letter awaiting him. It was from the Vice-Chancellor of Oxford University offer-ing him the newly created Rouse Ball Chair of Mathematics together with a fellowship at Wadham College at a salary of £1,200. (Breaking through £1,000 betokened affluence.) The appointment was his for the taking without any intimation that he might wish to go to Oxford and discuss it. Generous in death as in life, Rouse Ball endowed no fewer than three

---

[12] 8 Eldon Road (now Grove).

[13] Moon, P.B. 'Reminiscences and Discoveries' in *Not. and Rec. of the Roy. Soc. of London* 46, 1, (1992), pp. 171–174.

[14] Sir William Henry Bragg (1862–1942) held chairs at Adelaide, Leeds and University College London. Director of the Davy-Faraday Laboratory at the Royal Institution from 1923 to 1942. Nobel Laureate 1915.

[15] *Not. of Proc. of the Roy. Inst.* 25 (1926–1928), pp. 404–412. Reprinted in *Nature* 121 (9 and 16 June 1928), pp. 911–913 and 943–945.

professorships, two at Cambridge and one at Oxford, at £25,000 apiece, funded from his shrewd investments in the Guaranty Trust Company of New York that carried valuable tax advantages.

The offer could not have arrived at a more awkward moment. Besides the wedding and all that it entailed, Milne was on a tight schedule and had plenty on his mind. He had to squeeze in some external examining at Edinburgh before going to London for a big RAS meeting on 8 June. Many American astronomers would be there en route to the International Astronomical Union (IAU) conference at Leiden, and he was eager to talk to Henry Norris Russell and Edwin Hubble. At the RAS club dinner, swelled to record numbers by the guests, he sat opposite Sir Oliver Lodge,[16] pioneer of telecommunication, now more of a mystic than a physicist.

The disconcerting letter put him in turmoil. It dumbfounded him as much as it bewildered him. Only the previous autumn he had declined to be a candidate, although encouraged to apply, because he had become engaged to Margot and wanted to remain in Manchester. Of course an Oxford chair, with the lustre of the ancient university, had the trappings of advancement. On the other hand, amid his congenial colleagues at Manchester and on the threshold of marriage, his future shone bright with hope and promise. Moreover, he felt it was too soon to change jobs. He had barely completed three years and of this he had lost one term to his mastoid operation. As to practicalities, he had just incurred the considerable expense of buying, decorating and furnishing the Belfield Road house. Starting again would be a tiresome and expensive upheaval.

A contemporary article about Oxford professors consigned them to a sorry fate: 'No sooner are they elected to an Oxford chair than a blight descends on them.'[17] More to the point, Milne knew that he was not Oxford's first choice, and that other mathematicians did not want the chair. 'Fowler [would] not leave Cambridge and C.G. Darwin [did] not want to go to Oxford.'[18] Fearful of drawbacks that might jeopardise his

---

[16] Sir Oliver Joseph Lodge (1880–1932). First Principal of Birmingham University. After the death of his son Raymond in World War I he took up psychical research.

[17] 'Of Professors' in *The Oxford Magazine* 48 (21 November 1929), p. 224.

[18] Letter from Milne to A.V. Hill, 22 June 1928, AVHL I.

future wellbeing, Milne turned to A.V. Hill for advice and asked him whether there might be some 'snag'. Trusting him not to betray his confidence, Milne revealed that he nursed a secret aspiration to return to Cambridge: 'Newall has repeatedly urged me to be a candidate for the chair of astrophysics at Cambridge (which goes with the directorship of the SPO), when he retires. ... Newall regards me as his spiritual successor at the SPO.'[19] This was wishful thinking and lacked pragmatism. Besides having no appetite for administration, Milne was totally unsuited to supervising an experimental programme. His overriding desire was to pursue his own research, unfettered by the time-consuming obligations of running a department. It was true that the almost negligible demands of the Oxford chair dangled the carrot of greater freedom and, compared with Manchester, the city was cleaner and had a milder climate. Writing to Sir Joseph Thomson, Milne referred to the 'backwoods of Oxford',[20] but he seems to have had no inkling of the enormous edge Manchester had over Oxford with regard to science. Oxford's reputation lay in classics, theology and the humanities.

Whatever slight knowledge Milne had of Oxford probably stemmed from 1926, when he gave a paper on stellar atmospheres at a meeting of the British Association for the Advancement of Science. Most likely he watched the display of flying stunts over Port Meadow and the cricket match in the Parks organised by G.H. Hardy fielding a team of mathematicians. Certainly Milne heard Eddington's brilliant talk at the Oxford Union, which proved to be a career-defining moment for Milne's Brigand friend Jim Crowther, who was talent spotting for the Clarendon Press. Milne introduced Eddington to him, for which he was forever grateful, and Crowther persuaded Eddington to adapt his talk into a book. The result was the runaway best-seller *Stars and Atoms*, translated into ten languages. As well as setting a trend for popularising science, its success established Crowther's reputation and led to his appointment on the *Manchester Guardian* as Britain's first scientific journalist.

---

[19] Ibid.
[20] Letter from Milne to J.J. Thomson, 22 June 1928, papers of J.J. Thomson MS.Add.7654, Cambridge University Library.

This was hardly relevant to Milne's decision about the chair. No doubt he discussed it with Margot and the Fiddeses, but he seems to have failed to consider the matter in sufficient depth. Probably he was distracted by wedding preparations. He took notice that Eddington, whose opinion he revered, was one of the chair's electors. Persuasive letters from two other electors, G.H. Hardy and F.A. Lindemann, who knew Milne through the RAS, also carried weight in swaying him. Under their concerted pressure he accepted the chair a few days before the wedding. He warned Lindemann that he would fall short of his expectations in quantum mechanics as the 'hurly-burly of running the department at Manchester'[21] prevented his keeping up with the latest advances, and pointed out that the electors might have done better with Darwin, Fowler or Dirac.[22]

In this fateful and disastrous error of judgement Milne was blind to his own needs. He should have remained at Manchester. To function properly as a mathematician it was essential for him to have scientific companionship and informed criticism. This he blithely took for granted. Throughout the war and at Cambridge and Manchester he had basked in the warm approval that science and mathematics were worthwhile. He had no notion that at Oxford he would be cast adrift in an environment that did not put a high value on science or on research.

On 26 June 1928 the Reverend Alexander Fiddes married my parents in the Presbyterian Church at Withington. Edward and Agnes Fiddes spared nothing in providing a splendid reception for family, friends and colleagues. Margot had a page and three bridesmaids; Geoffrey was best man, accompanied by James Duff, renowned for his eloquent wit and now a lecturer in education. The newly-weds had a week's honeymoon touring the Cotswolds, through Tewkesbury, Cheltenham and Bibury, finishing at Oxford. They went on botanising walks and afterwards, in her copy of Bentham and Hooker's *The Flora of the British Isles*, Margot coloured in the black and white illustrations of the flowers they had identified.

---

[21] Letter from Milne to F.A. Lindemann, 15 June 1928, Cherwell papers, Nuffield College, Oxford.

[22] Paul Adrien Maurice Dirac (1902–1984). Quantum physicist. Lucasian Professor at Cambridge from 1932 to 1969. Nobel Laureate 1933. He refused Manchester's Beyer Chair. See Kragh, H. *Dirac*, Cambridge: Cambridge University Press, 1990, p. 60.

Milne was always in a hurry. They sped back to Manchester for a quick two-day turn-around at their Belfield Road house before joining the IAU meetings in Holland. The conference was almost twice the size of the previous one in Cambridge and the first since the war to have German astronomers present. At the opening ceremony in The Hague, the flags around the room testified to the international character of the meetings. Margot was able to meet Milne's wide circle of acquaintance from Britain, Europe and the United States, and she went on the programme for spouses while he was busy with IAU commissions (on solar physics, stellar spectra and stellar constitution). Together they enjoyed the canal excursions for all participants.

Milne postponed starting at Oxford until 1 January 1929 so that he and Margot could have a little time in the house they had prepared. They put it to good use and among a string of house guests were Edith Milne, Hugh Campbell, Edwin Rayner from Teddington and John Littlewood, Cambridge's new Rouse Ball Professor. That autumn Philip began his degree course in Manchester, financed by his brother. Milne had hoped that Philip would study physics, but his mathematics was not up to it and he became a zoologist.

Milne's halcyon Manchester days brimmed with professional and personal fulfilment. He had gained not only a wife but a most congenial new family. He was an accepted force in astronomy and a Fellow of the Royal Society. His departure signified a watershed in his research. A year later he veered in a fresh direction and put forward ideas of such extreme flights of fancy that they stretched credulity.

Chapter 8

A Scientific Wilderness

An Oxford wit defined science as a continuous discovery of its own mistakes.[1]

Milne was right to be apprehensive about Oxford. Even before he arrived, his college, Wadham (Fig. 10), rebuffed him. While he and Margot were in Oxford looking for somewhere to live he called on the Warden of Wadham, John Stenning,[2] an Aramaic scholar, to ask him for a room in college where he could hold colloquia. Milne was keen to instigate them, as he had been impressed by their value at Manchester. Aware that mathematics professors at Cambridge had rooms in the Arts School, and that Oxford had no similar provision, nor a departmental centre for mathematics, he thought his request entirely reasonable. He was mistaken. The Warden admitted that Milne was entitled to a room, but, sweeping his request aside, said that as a married man he would not 'want' one, and deliberately changed the conversation.

For a charitable explanation of the Warden's brush-off, it is worth recalling that Milne's Wadham fellowship was linked to his Rouse Ball Chair by a university statute which tied each professorial chair to a named college. The arrangement was not without its critics. The Fellows had scant say, if any, in selecting the professor, and might resent having him foisted on them, particularly in the case of a small college, such as Wadham. Besides, if the new professor came from outside Oxford, he might be perceived as intruding into their intimate circle. Almost certainly he would be the most junior Fellow, and, by custom, expected to hand round the nuts and sweetmeats in the senior common room after dinner.

---

[1] Ivor Brown, *The Times* (18 January 1933).
[2] John Frederick Stenning (1868–1959). Warden from 1927 to 1938.

103

The Warden's decision was symptomatic of the university's sloth to modernise and its outdated, stuffy attitude. It was, for instance, only two years previously that the university had abandoned 'divvers', the scripture examination compulsory for a BA degree. In 1929, when readers at the Bodleian Library enjoyed electric light for the first time, the Oxford University Observatory (OUO) had not yet converted its electricity supply from direct to alternating current. From the Warden's viewpoint, college rooms were for bachelor dons who devoted their undivided attention to their pupils and to college affairs. College loyalty and college teaching came first. The hallmarks of a seasoned Oxford don, admired by his peers, were a succession of brilliant pupils who outshone those in rival colleges, and a reputation for dazzling repartee laced with clever aphorisms.

Vexed and taken aback by the humiliating snub, Milne was determined to get his way, but, equally, he was anxious not to cause offence at this early stage. He needed an ally. Hardy was the obvious choice but he was away in Princeton, and so Milne turned to Lindemann, who was a Professorial Fellow of Wadham, although he did not reside there, preferring the better plumbing and grander comforts of Christ Church, to which his chair was originally assigned (until Christ Church protested that it did not want to have a scientist on its governing body). Tall, imposing, wealthy, possessed of a mordant wit and a withering tongue, Lindemann could exert hefty clout and he bullied the Warden into capitulating. Milne made excellent use of his room and his innovative weekly colloquia were long remembered.

When Hardy came to hear of Milne's treatment he was disgusted: 'Why should a professor have to grovel before a college bursar, when he wants a room and a decent blackboard?'[3] Yet in luring Milne to Oxford, Hardy had been economical with the truth. He well knew of the deficiencies that beset an Oxford mathematician: 'The first thing which a Cambridge man notices when he migrates to Oxford is that mathematics and physics, the "ranking" subjects in his old university, are overshadowed in Oxford not merely by the literary schools, but even by other sciences.'[4] Chemistry was ever the leading physical science. Oxford ran

---

[3] Hardy, G.H. 'Mathematics' in *The Oxford Magazine* 48 (5 June 1930), p. 821.
[4] Hardy, G.H. 'Mathematics' in *The Oxford Magazine* 48 (5 June 1930), p. 819.

fewer courses in mathematics, and pupils were not comparable 'either in numbers or quality to those taught at Cambridge'.[5] To give Hardy his due, he probably had the best of intentions and expected that Milne would upgrade Oxford's mathematics. However, when Hardy had the chance, in 1931, he returned to Cambridge for good.

Implausible as it seems, it used to be the practice for Oxford to pack its most promising mathematics students off to Cambridge for special coaching before their final examinations![6] This astonishing arrangement reflected the shortage at Oxford of dons who taught mathematics; almost half the colleges, including Wadham, did not have a tutor in mathematics. Such colleges had no choice but to farm out their undergraduates reading mathematics to other colleges for instruction.

A factor which militated against science and mathematics candidates applying to Oxford was the rife snobbery. To an admissions tutor, seeking the most promising undergraduates, a well-spoken applicant in classics from a famous school seemed a more attractive proposition than a North Country youth applying for science or mathematics. A.P. Herbert, the epitome of an Oxford man, remarked that 'we did not ... take science very seriously'.[7] Its inherent specialisation ran counter to the image of a successful Oxford graduate grounded in the humanities and able to advance a cogent argument with éclat. Theology, law and history dominated. Oxford bred bishops, judges, ambassadors, cabinet ministers, men of letters and those who rose to eminence in the civil, colonial and foreign services. It was assumed that scientists were incapable of acquiring the skills which Oxford most prided because, immersed in the minutiae of their chosen fields, they lacked the grooming for the traits which the university held most dear.

The nature of scientific enquiry was generally misunderstood. One wag claimed that a man who had taken a First in 'greats' (classical studies) could master physics in a fortnight.[8] Arts dons did not accept that

---

[5] Letter from G.H. Hardy to J.J. Thomson, 16 December, no year, Thomson papers, Cambridge University Library. Oxford had no more than twenty-five finalists a year.
[6] Fauvel, J., Flood, R. and Wilson, R. (eds). *Oxford Figures*, Oxford: Oxford University Press, 1999, p. 26.
[7] Herbert, A.P. *A.P.H., His Life and Times*, London: William Heinemann, 1970, p. 21.
[8] Ayer, A.J. *Part of my Life*, London: Collins, 1977, p. 143.

scientists, like themselves, thrived on inspiration and intuition and were just as vulnerable to human fallibility and judgement.

Milne was dismayed to be relegated to the backwaters of the university simply for being a mathematician. Even more galling was his discovery that original scholarship counted for little, and furthermore that a professor was virtually powerless with no funds at his disposal. To compound Milne's disenchantment, he found Oxford to be inward looking. He was shocked by the bickering between members of the university, and contrasted it with the mutual admiration that was indigenous to Trinity and conducive to research. J.J. Thomson believed that research brought maturity of mind,[9] while Oxford suffered from a feeling that all things had been discovered, and scorned research for being an ungentlemanly pursuit. After completing a BA degree, an aspiring graduate was unlikely to embark on research but would read for a second BA degree in another subject to broaden his base for teaching. The crowning accolade was the trophy of a coveted university prize.

It was unfortunate that when Milne joined the university its science was at a nadir. Although Lindemann had resuscitated research at the moribund Clarendon Laboratory, his animosity with John Townsend,[10] the leading authority on ionised gases, for whom the Drapers Company funded the Electrical Laboratory, fractured the study of physics. Prejudice against the sciences was fuelled by a hostile climate of distrust that whipped up suspicion that science was expanding and encroaching on traditional academic disciplines. Science subverted college loyalty and it was feared that the cost of new laboratories would gobble up university finances.

By association, mathematics was a victim of this prejudice because it was a sub-faculty[11] of the Board of Physical Sciences. Since mathematics is indispensable to science, it was regarded as its handmaiden, although since ancient times it has been an independent subject in its own right. Far from gobbling up finances, it is inexpensive to teach, requires few textbooks and has no laboratory expenses. To its detriment, Oxford mathematics had no

---

[9]Eve, A.S. *Rutherford*, Cambridge: Cambridge University Press, 1939, p. 9.
[10]Sir John Sealy Edward Townsend (1868–1957). Irish physicist. He was appointed to the Wykeham Chair in 1900.
[11]In 1963 mathematics attained the status of an independent faculty.

formal links with philosophy, a subject of great prestige within the university, whereas at Cambridge Bertrand Russell had irrevocably drawn them together.

The insistence by the colleges on preserving their autonomy and authority accelerated the dismal decline in science. They fiercely opposed the sensible proposals of the 1922 Asquith Committee that advised centralising science teaching in university laboratories and wresting control away from colleges, which insisted on clinging to the tutorial system that suited the humanities. College chemistry laboratories, often unhealthy, usually dingy, continued to function. This was in marked contrast to Cambridge where the Cavendish Laboratory even accepted people from overseas, such as Rutherford.

Things had not always been like this. In bygone days Oxford was a major centre of science. In the fourteenth century its latitude was the navigational reference for the Western world. In the seventeenth century Wadham was at the heart of scientific endeavour. Christopher Wren mounted his telescope in the tower above the main gate. Warden John Wilkins, friend of John Evelyn and brother-in-law of Oliver Cromwell, was well ahead of his age and envisaged flying to the Moon in a chariot. In his rooms, meetings with Edmund Halley, Robert Boyle and his assistant Robert Hooke, germinated the seeds of the Royal Society. Moreover, Oxford was the first English university to open a science museum for the public, the old Ashmolean, and to endow a chair in astronomy.

Beyond the dispiriting revelation that mathematics and research were now relegated to the margins, Milne had personal grounds for disappointment. His ingrained sense of duty demanded he cultivate allegiance towards his new university, not replacing one loyalty by another, but expanding them. He was peeved that far from extending a welcome, the university put a stumbling block in his way. 'Whilst every individual is extremely friendly,' he told Lindemann, 'the University itself is an unfriendly institution. It starts its career of unfriendliness towards an incoming professor by charging him degree fees.'[12] To become a member of the university, Milne was required to acquire an Oxford MA degree, and — this was the rub — he was expected

---

[12]Letter from Milne to F.A. Lindemann, 12 December 1931, Cherwell papers, Nuffield College, Oxford.

to pay for the privilege. Possessing the MA was no mere formality. Without it he would be ineligible to serve on the myriad of committees which controlled appointments, managed buildings and so forth. The Nobel Laureate Frederick Soddy, who coined the word 'isotope', was so outraged that the university did not grant him an MA out of courtesy that he doggedly refused to pay, and in consequence debarred himself from taking part in any university business.

Milne was determined to elevate mathematics out of its stagnation and to obtain departmental premises. Although he was no lover of administration he paid for his MA and was soon serving on several committees. But not one committee, he was astounded to find, convened all thirteen of the professors in science and mathematics, with the ineluctable consequence that it was impossible to formulate cohesive policy, a situation he deplored. He missed a sense of common purpose.

The unofficial nerve centre for deliberating policy was a select dining club, the Oxford University Natural Science Club, to which entry was by election. Twice a term, over a good dinner, its twenty members discussed matters ranging from examination syllabuses to discrepancies in stipends. (Those who engaged exclusively in research came off worst of all as they did not earn fees from teaching.) As soon as Milne gained election to the club, in 1932, he added his voice in petitioning the Hebdomadal Council, the university's highest authority, to increase the number of scientists on its body. No scientist had a hand in running university finances and, as yet, no scientist was head of a college.

Milne was one of four professors of mathematics at Oxford. Hardy had brought fresh impetus, but the other two[13] had long ceased to make any impact. Like Herbert Turner,[14] the Professor of Astronomy, they were in their sixties and appointed under old statutes that bestowed the sinecure of life tenure on full salary. In these circumstances Milne, young and bubbling with vivacity, had ample scope for injecting enthusiasm. He was the obvious choice for graduate students choosing a research supervisor, and to have him students were willing to switch from pure to applied mathematics, or even to change subjects.

---

[13] Augustus Edward Hough Love (1863–1940) was appointed to the Sedleian Chair in 1898. Arthur Lee Dixon (1867–1955) was appointed to the Savilian Chair in 1922.

[14] Herbert Hall Turner (1861–1930) was appointed Savilian Professor of Astronomy in 1893.

Few professors belonged to the Oxford University Mathematical and Physical Society, but Milne went out of his way to support it. Despite its name, the society did not straddle areas of mathematics and physics that overlap, which would have appealed to him, but held alternate meetings in the separate disciplines. Milne took guests, encouraged membership, enlivened its proceedings by sparking impassioned discussion and became president.

During these first months, while he was trying to adjust to his disillusion, Milne was preparing a prestigious lecture for the Royal Society. The Bakerian, endowed by Henry Baker, an expert in microscopy and son-in-law of Daniel Defoe, is the Royal Society's premier lecture in the physical sciences. Acutely conscious of the honour of joining an illustrious line of past lecturers, including Thomson, Rutherford and Eddington, Milne was keyed up to do his best. He was of a mind to unveil his latest research which explored the physical conditions within a star, undeterred by the fact that he was a novice about its internal structure.

At Oxford Milne worked in a vacuum and desperately needed someone like Ralph Fowler, with whom he could engage in meaningful discussion and who could check the excesses of his fertile brain. Obsessed by his belief in the overriding power of mathematics without paying due regard to the physics, Milne was surprised to deduce that the centre of a star is extremely hot and dense. This conclusion disagreed with Eddington's theory that stars are perfect gases. Clearly Milne needed to consult Eddington — whose judgement he valued — as he had often done in the past.

At this juncture accounts diverge. One version has it that Milne went to see Eddington in Cambridge, and was on the brink of launching into a detailed exposition of his ideas when the latter indicated he wanted quiet, and stared pensively out of his study window. He had an extraordinary gift of insight, and after a few moments he broke the silence by taking Milne's words out of his mouth. Amazed, Milne agreed that that was precisely what he was going to say. 'Well,' said Eddington, 'don't.'[15] Milne's account is less dramatic:

In April 1929, after working for 3 months on new ideas, I found my work beginning to run (unexpected by me) counter to Eddington's, so I sent it all to him to see what he thought. He sent it back saying it was obviously

---

[15] Conversation with Sir William McCrea, 5 February 1990.

wrong, though I could hardly expect him to put his finger on the mistake. However, I was to look in such and such a place. After 3 or 4 letters on each side I (believing I must be wrong) thought I detected a mistake and withdrew the whole thing. This was 4 weeks before my Bakerian lecture, and I found myself with all my preparations turned to dust and ashes.[16]

Whichever version is closer to the truth, Eddington's veto demolished Milne's cherished plans at a stroke. With hindsight it was a blessing though, because it saved Milne from giving an ignominious lecture and, as time would show, Eddington's intuition was vindicated.

Milne was now in a tight spot. At this late stage he had to rack his brains to compose from scratch another lecture worthy of the occasion. Wisely confining himself to his old stomping ground of stellar atmospheres, Milne investigated the Sun's photosphere. The lecture was a great success and was rated his most useful work since establishing stellar temperature scales with Fowler, though, as Milne pointed out, his analysis would need modifying if hydrogen predominated, as later proved to be the case.

In the rush to meet the deadline Margot carried out arithmetical calculations and Milne enlisted the help of his first Oxford research student to compile tables of data. Tom Cowling,[17] a spare, tall young man with ginger hair, who would make his name in solar magnetism, claimed he was an astronomer by accident,[18] an accident triggered by Milne. Cowling had gained a first-class degree in mathematics, but, because he deemed the course out-dated, decided to change direction and undertake a doctorate in atomic physics. He consulted his tutor, Idwal Griffith,[19] who was unusual in that he was both a mathematician and an experimental physicist. was also a cheerful and prominent university 'pooh-bah'[20] and he

---

[16]Letter from Milne to H. Dingle, 1 August 1930, EAM, MS.Eng.misc.b.429.

[17]Thomas George Cowling (1906–1980). Astrophysicist. Professor at Leeds University from 1948 to 1970.

[18]Cowling, T.G. 'Astronomer by Accident' in *Am. Rev. Astron. Astrophys.* 23 (1985), pp. 1–18.

[19]Idwal Owain Griffith (1880–1941). Fellow of St John's, then Brasenose.

[20]Attributed to A.H. Cooke, Warden of New College, in Croft, A.J. *Oxford's Clarendon Laboratory*, Oxford: [s.n.], (1986), p. 77. A typescript copy is held by the Radcliffe Science Library at shelfmark RR.x. 384.

wielded influence on many committees. He had backed Milne's appointment from personal knowledge after striking up a friendship with him at Orfordness where, a major in the RFC, he had worked on the aeronautics of night flying. His advice to Cowling was to write to Milne at Manchester and try to secure him as his supervisor. Milne replied that his atomic physics was rusty but agreed to take Cowling on if he were willing to change to astronomy, and enclosed a challenging reading list.

In January, when the Milnes arrived, Oxford was in the grip of a severe winter. The intense cold of the unheated old Clarendon Laboratory, where Milne lectured, tested Cowling's concentration. Even so, he thought Milne's course on stellar atmospheres a 'masterpiece',[21] building up theory bit by bit from fundamentals. Milne made no concessions to the weaker students, who found the going hard, but his infectious enthusiasm gave the impression that 'he had a mathematical secret which he was desperate to share. There was an air of excitement, creating in us a feeling that "This may be good, but it is going to get better".'[22] By the end of each hour Milne had exhausted his listeners, leaving them as breathless as he was.

Milne had no serious competitors as a lecturer. His diction was clear, and he was a 'prince'[23] compared with Lindemann's inaudible monotone. One mathematics lecturer was so inept that the audience called out what he should write down. Another worked backwards. Having started on one proof he realised it relied on an earlier theorem, so began on that, and so on.

Cowling had only enough funds for a two-year DPhil. Given this restriction, Milne had difficulty thinking of a topic pitched at the right level and got in touch with Sydney Chapman, who outlined a problem about the Sun's magnetic field. Six months later, after Cowling had made scant progress, Milne reminded him of his original intention of taking up a career in teaching. Eventually, though, his research blossomed and it led to a fruitful collaboration with Chapman with whom Cowling wrote a textbook.

---

[21] Interview with T.G. Cowling by D. DeVorkin, 22 March 1978, Niels Bohr Library and Archives, American Institute of Physics, College Park, MD, USA.
[22] Letter from G.L. Camm to the author, 5 October 1989.
[23] Letter from E. Norris to the author, 2 April 1991.

Just as Cowling began to make headway, my parents left for a liberating trip to America. Milne was to lecture at Ann Arbor's summer school, in Michigan, and for six blissful weeks he would be among like-minded souls with whom he could exchange ideas. Afterwards, with a £30 grant from the Royal Society, he was to have a month at the Harvard Observatory working with Cecilia Payne,[24] his first Cambridge pupil, on her incomparable collection of spectroscopic data. The summer promised cathartic relief from Oxford's scientific wilderness.

Milne was keen to reinforce his links with American astronomers. With their clear skies and well-funded equipment, they led the world in observational astronomy, whereas British astronomers excelled in theoretical astronomy. Co-operation was the essence of the rapid advances made in astrophysics, and Milne's visit was opportune. An earlier possibility of visiting the USA had arisen when Milne leapt at the chance to talk to George Ellery Hale when he was passing through London. In frail health, Hale summoned Milne to his hotel and aired the attractive prospect of him spending some time at the Mount Wilson Observatory.[25]

Milne hoped to confer with Henry Norris Russell, but, ever a hypochondriac, he was on an extended rest in Europe. Shortly after arriving in Oxford and released from running a department, Milne had used his new-found freedom to study Russell's influential two-volume *Astronomy*, co-written with his Princeton colleagues, R.S. Dugan and J.Q. Stewart.[26] The last chapter on the evolution of stars had sparked Milne's curiosity, and he started to speculate about their internal structure. Despite Eddington's warning, Milne continued to brood on the subject.

The Milnes crossed the Atlantic in a luxurious White Star liner, docked at Boston and then caught a train to Chicago, where they went up the Tribune Tower. At Ann Arbor, the friendly director of its observatory,

---

[24]Cecilia Helena Payne Gaposchkin (1900–1979). In 1956 she became the first woman to be a full professor at Harvard.

[25]Letters from Milne to G.E. Hale, 23 April 1929 and 20 January 1932, Hale papers, California Institute of Technology.

[26]In 1905 Raymond Smith Dugan (1878–1940) joined the staff at Princeton, followed in 1921 by John Quincy Stewart (1894–1972).

Ralph Curtiss,[27] a noted authority on stellar spectra, gave the Milnes a warm welcome. During their stay, they enjoyed, along with other participants, a steamer excursion on Lake Erie and cavorting in the spray at Niagara Falls, dressed in waterproofs.

The summer school was the brainchild of Harrison Randall,[28] who had built up an outstanding Physics Department. During a sabbatical leave in Germany, he had been struck by the advantages of combining theoretical and experimental physics within a single department, a practice then unknown in America. Wishing to emulate the German system, he recruited the Swedish atomic physicist Oscar Klein,[29] and the discoverers of electron spin, the Dutchmen Samuel Goudsmit[30] and George Uhlenbeck.[31] Randall instituted summer symposia on theoretical physics, which he enriched with guest speakers, and these escalated into the renowned summer schools. When participants returned to their universities they disseminated what they had learnt about the latest developments in quantum physics.

The theme of the 1929 summer school was the application of quantum theory to stellar atmospheres. Besides Milne, Randall imported Paul Dirac, whose relativistic equation for the electron reconciled quantum physics with special relativity; the Frenchman Léon Brillouin;[32] and the Austrian-born Karl Herzfeld.[33] Milne gave a course on Sunspots and lectured at such a pace that one listener vowed he had never heard so much covered in an hour.

---

[27]Ralph Hamilton Curtiss (1880–1929). In 1907, after some years at the Lick Observatory, he moved to Ann Arbor. Milne's obituary of him is in *The Observatory* (February 1930), pp. 54–56.

[28]Harrison McAllister Randall (1870–1969). Professor at Ann Arbor from 1905 to 1940.

[29]Oscar Benjamin Klein (1894–1977). Professor at Ann Arbor from 1923 to 1925 and at Stockholm from 1938 to 1960.

[30]Samuel Abraham Goudsmit (1902–1978). Professor at Ann Arbor from 1927 to 1946. From 1948 to 1970 he held positions at the Brookhaven National Laboratory, Long Island.

[31]George Eugene Uhlenbeck (1900–1988). Apart from a few years elsewhere, he held positions at Ann Arbor from 1927 to 1960. From 1960 until 1974 he worked at the Rockefeller Research Center in New York.

[32]Léon Nicholas Brillouin (1889–1969) emigrated to the USA in 1941 and held appointments at Wisconsin-Madison, Brown, Columbia and Harvard.

[33]Karl Ferdinand Herzfeld (1892–1978). Professor at Johns Hopkins University from 1926 to 1936, and at the Catholic University of America, Washington DC, from 1936 to 1978.

Among his audience, assiduously taking notes, were two astronomers whom Milne knew by reputation: an American, Theodore Dunham,[34] aged thirty-one, from the Mount Wilson Observatory; and a Canadian, Harry Plaskett,[35] aged thirty-six, from the Harvard Observatory. (Plaskett's work on high surface temperature O-type stars had sharpened Milne's interest in stellar ionisation.) They and their wives were lodging at the same fraternity house as the Milnes. The three couples got on famously and their friendship was to shape future events. In the heat of high summer the wives cooled off by swimming in the lake and the husbands indulged in 'ice cream parties', vying to eat the most ten-cent dishes at their favourite drug store while they talked and talked. They shared a fixation with perfection, and nothing was too trivial to be left to chance.

Plaskett, who cited Milne as the 'star'[36] of the summer school, was primarily an observer, Dunham's forte was optical instruments, and Milne was a theoretician, so each brought his own perspective. Dunham's and Plaskett's paths to astronomy were radically different. Plaskett grew up with astronomy in his veins. His father, John Plaskett,[37] whom Milne had met, was Canada's foremost astronomer, Director of the Dominion Astrophysical Observatory in Victoria, British Columbia, which was funded and constructed by the government to his specification. Harry Plaskett had taken a wartime degree at the University of Toronto before fighting at Passchendaele. Under a scheme for Canadian servicemen he studied in London at Imperial College with the renowned spectroscopist Alfred Fowler,[38] an authority on Sunspots. Returning to Canada, he used the 72" reflecting telescope in his father's observatory to measure the amount of helium and hydrogen in

---

[34] Theodore Dunham (1897–1984). Physician and astronomer. He joined the staff of Mount Wilson Observatory in 1928. From 1948 to 1957 he worked at the University of Rochester, then moved to Australia to build the spectrograph at Mount Stromlo Observatory and remained there, latterly in Tasmania, until 1970.

[35] Harry Hemley Plaskett (1893–1980). Solar physicist. Savilian Professor of Astronomy at Oxford from 1932 to 1960.

[36] Letter from H.H. Plaskett to N. Hetherington, 25 June 1978, in the author's possession.

[37] John Stanley Plaskett (1865–1941) became Director of the Dominion Observatory in 1917.

[38] Alfred Fowler (1868–1940). Professor at Imperial College London.

Fig. 1. Arthur, left, and Geoffrey, c.1908.

Fig. 2. Great Court, courtesy of the Master and Fellows of Trinity College.

Fig. 3. Sir Horace Darwin, courtesy of Horace Barlow.

Fig. 4. Captain A.V. Hill with his daughters Polly and Janet, courtesy of Nicholas Humphrey.

Fig. 5. Farman Experimemtal 2b pusher plane, courtesy of the Shuttleworth Collection, Biggleswade.

Fig. 6. Lieutenant Milne RNVR.

Fig. 7. Men operating listening trumpets, Q 35879, by permission of the Imperial War Museum, London.

Fig. 8. G.H. Hardy, courtesy of the Master and Fellows of Trinity College.

Fig. 9 Mathematics graduates, Manchester, 1926. Gift of George Tyson.

Fig. 10.   Wadham College, Oxford, courtesy of the Warden and Fellows.

Fig. 11.   Theodore and Miriam Dunham, Margot and Arthur Milne, William G. Thompson, Mount Washington, New Hampshire, 1929.

Fig. 12. Chandra, Chicago, 1985.

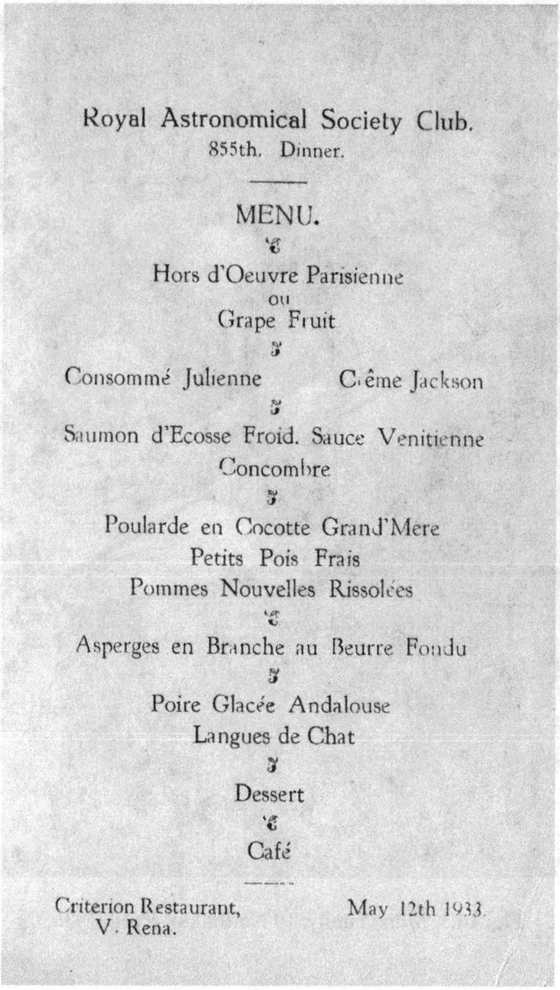

Royal Astronomical Society Club.
855th. Dinner.

MENU.

Hors d'Oeuvre Parisienne
ou
Grape Fruit

Consommé Julienne   Crême Jackson

Saumon d'Ecosse Froid. Sauce Venitienne
Concombre

Poularde en Cocotte Grand'Mere
Petits Pois Frais
Pommes Nouvelles Rissolées

Asperges en Branche au Beurre Fondu

Poire Glacée Andalouse
Langues de Chat

Dessert

Café

Criterion Restaurant,   May 12th 1933.
     V. Rena.

Fig. 13.   Menu card for RAS club dinner, 12 May 1933.

Fig. 14. Milne family, 19 Northmoor Road, 1936.

Fig. 15. The Einsteinturm, Potsdam, and the author, 2010.

Fig. 16. Back row from left: Arthur Milne, Harry Plaskett, Henry Norris Russell, A.E.H. Love. Front row from left: Margot Milne, Lucy Russell, Edith Plaskett. Taken by Thornton L. Page, gift of R. Hutchins and reproduced by his kind permission.

Fig. 17. Russell and Milne in conversation, 1939.

Fig. 18. Milne, taken by Chandra, May 1939.

Fig. 19. Wedding day, 22 June 1940.

Fig. 20. Milne's medals: Bruce Medal, Royal Medal of the Royal Society, Gold Medal of the Royal Astronomical Society, Johnson Memorial Prize.

Fig. 21. Milne family, Worthing, November 1945.

stellar atmospheres. At the Harvard Observatory he set up a graduate programme in astronomy, the first of its kind in the USA.

Sorrowful circumstances had brought the Dunhams to Milne's attention,
as they had shown great kindness to a young friend of his, Sydney Pike.[39]
An Oxford physicist of exceptional promise with an attractive personality,
Pike had contributed to the study of solar prominences and Sunspots and
won a fellowship to Mount Wilson Observatory. Shortly after arriving he
fell unaccountably ill with osteomyelitis of the skull. Theodore and Miriam
Dunham got him the best possible medical care but Pike died, aged twenty-
five. They gave hospitality to his fiancée, Ruth Garstang, who came out
from England, and witnessed with her the scattering of Pike's ashes on the
slopes of Mount Wilson below the observatory buildings.

Dunham came from a distinguished line of medical doctors, and, unlike
Harry Plaskett, his entry into astronomy was a triumph of patience and
perseverance. Since boyhood the stars had fascinated him. He read astronomy books borrowed from the New York Public Library near his home,
saved up his pocket money to buy a telescope and fixed up a little observatory at his family's holiday house in Maine. He yearned to be an astronomer but his autocratic father doubted whether he could earn a living in
astronomy and insisted that after graduating from Harvard he go to medical school. Theodore obediently qualified at Cornell, coming top of his
year. Then, having proved himself, he astonished his family by renounc-
ing medicine. He approached Henry Norris Russell, who was bowled over
by Dunham's self-taught grasp of astronomy and who accepted him forthwith for a doctorate at Princeton. Russell tossed him a roll of film of the
spectrum of the star α-Persei, with the words, 'Here's your thesis.'
Estranged from his family, Dunham was grateful to Russell for his patronage. He ensured that Dunham had enough money to live on and after his
doctorate recommended him for a position at Mount Wilson Observatory,
then in its heyday. When Milne met him, Dunham was assisting the
director, Walter Adams[40] to design and install the Coudé spectrograph for

---

[39] Sydney Royston Pike (1903–1928). Milne's obituaries of him are in *Mon. Not. RAS* 89
(1929), pp. 319–320 and *The Observatory* (December 1928), pp. 381–382.
[40] Walter Sydney Adams (1876–1956). He succeeded George Ellery Hale as Director of the
Mount Wilson Observatory in 1923.

the 100″ Hooker telescope. In 1932 they identified carbon dioxide in the atmosphere of Venus. Dunham's meticulous attention to detail was legendary. Merely to fetch exactly the right spanner for a particular job he would dash down the dirt mountain road to the workshops in Santa Barbara Street, Pasadena, and up again.

Whenever Russell visited Mount Wilson, he took the Dunhams for a weekend camping trip in the desert, providing little more than cartons of eggs for nourishment. He expected them to swallow a raw egg in a single gulp. Russell, restless as ever, never stopped imparting his encyclopaedic knowledge. By day he identified the desert wildflowers, and by night, as they lay on their camp beds gazing upwards, he reeled off the characteristics of each star as if it were a personal friend. On Sunday morning they stood in worship while Russell read the service from the Presbyterian prayer book and the Dunhams, who were Episcopalians, made the responses.

The dynamics of the Dunham/Milne friendship gained momentum. Milne liked bouncing ideas off Dunham and supplied him with a computation in optics that he needed. Dunham identified Milne as a 'penetrating quick thinker, quiet, but cheerful and companionable'.[41] He told his wife, 'The Universe is giving up its secrets … [Milne] revealed a little of his soul to me … it is all about a star and how it keeps going. … We get on better and better.'[42]

Their growing closeness induced Miriam Dunham to invite the Milnes to come to her parents' house outside Boston and to their vacation retreat in New Hampshire. Privately she hoped that Milne's intellectual vigour would be a consoling diversion to her father, William G. Thompson, a highly esteemed senior member of the Boston bar, who was under a demeaning and undeserved professional cloud. He had defended the Italian immigrants and anarchists Nicola Sacco and Bartolomeo Vanzetti,[43] one a shoemaker, the other a fish pedlar, who were found guilty of murder.

---

[41] Letter from M.P. Dunham to the author, 10 February, no year.

[42] Letter from T. Dunham to M.P. Dunham, 1929, typed copy in the author's possession.

[43] In 1977 the Governor of Massachusetts issued a proclamation which affirmed that the trial was permeated with prejudice and removed any stigma and disgrace attached to the men.

Their notorious trial, which lasted six years, reverberated with public outcry and caused mass demonstrations. Upton Sinclair wrote in his novel *Boston*, which reminded Milne of Galsworthy's *The Forsyte Saga*,[44] that Thompson, dry and humorous, was 'the boss of which the case had been in need'.[45] In his 'brave, learned fight'[46] — when money ran out he worked without a fee — he argued that class and political prejudice denied the Italians a fair trial. Appeals for clemency to the Governor of Massachusetts and the President of Harvard went unheeded. On the eve of his electrocution, Vanzetti asked Thompson to come to his cell. He entreated Vanzetti to forgive his enemies, 'In the long run the force to which the world will respond is the force of love and not of hate.'[47] Thompson was ostracised for his highly principled and uncompromising stand and incurred the calumny of die-hard conservatives who distanced themselves from him. His law practice declined and he no longer felt able to eat lunch at his club.

Thompson had a sound understanding of science, and, as Miriam anticipated, he and Milne took to one another. Milne was impressed by his astute questions. In their tranquil house on the brow of a hill, where the magnificent view takes the eye across the valley to the peaks of the White Mountains, the Thompsons entertained the Milnes with unobtrusive generosity. The week glided by in picnics, tennis, walks in the pinewoods, a hike up Mount Whittier and a drive to the summit of Mount Washington (Fig. 11).

Next stop was Cambridge, Massachusetts. Earlier, while in Boston, Milne had called on Cecilia Payne at her office in the Harvard Observatory. She was nervous lest Milne should accidentally encounter another visitor, Albrecht Unsöld,[48] who happened to be in the building, because they were in the throes of a heated controversy concerning the chromosphere. Suddenly Unsöld appeared at the door. 'They stood there: Milne small,

---

[44]Letter from Milne to W.G. Thompson, 16 August 1930, copy in the author's possession.
[45]Sinclair, U. *Boston*, London: Werner Laurie Ltd, 1929, p. 502, reprinted by Robert Bentley, 1978.
[46]'Sacco and Vanzetti' in *The Oxford Magazine* (16 May 1929), pp. 615–616.
[47]Thompson, W.G. 'Vanzetti's Last Statement' in *Atlantic Monthly* (February 1928), pp. 254–257.
[48]Albrecht Otto Johannes Unsöld (1905–1995). Director of the Institute of Theoretical Physics at the University of Kiel from 1932 to 1973.

delicate, fair; Unsöld tall, dark, forbidding. There was a long silence. Then someone began to laugh; the gesture was contagious, and soon we were all laughing.'[49]

Ten years previously when Cecilia Payne was an undergraduate at Newnham College, Cambridge, she was swept away by Eddington's historic talk about the eclipse, and immediately resolved to become an astronomer, despite her lack of mathematics. Suspecting that as a woman she would be unlikely to forge a career in astrophysics in England, she emigrated to America, but even at the Harvard Observatory, formidable as she was, she laboured away for many years in a lowly capacity. The theory of stellar thermal ionisation was the springhead of all her early work. Applying Fowler's and Milne's analysis to the observatory's vast collection of plates made her name. She confirmed that hydrogen was the predominating element in stars and after she battled widespread scepticism, this was accepted.

Her spectroscopic data held the key to a huge reservoir of stellar information, and the prime question in Milne's mind was how best to interpret it. He wanted to calculate the opacity of a stellar atmosphere, the depth at which an absorption line originates, and if time allowed, to deduce stellar masses, temperatures and chemical composition. He would willingly have stayed longer among the amiable Harvard astronomers but knew it would not be acceptable to miss the start of his first full Oxford academic year. During the precious month he contributed a paper to the observatory's *Bulletin*, and gave 'beautifully clear modest'[50] lectures on the chromosphere.

The director of the observatory was Harlow Shapley,[51] whose estimate of the size of our Galaxy showed that our Solar System is not at its centre. He and his wife Martha liked the game of charades, and Milne landed the part of the Reverend Jeremiah Horrocks, 'faithfully combining observations of Venus with religious duties'.[52] At the remarkable age of twenty,

---

[49]Haramundanis, K. (ed). *Cecilia Payne Gaposchkin*, Cambridge: Cambridge University Press, 1996, p. 179.

[50]Letter from H. Shapley to H.N. Russell, 3 September 1929, Shapley papers, Harvard University Archives.

[51]Harlow Shapley (1885–1972). Director of the Harvard Observatory from 1921 to 1952.

[52]Letter from H. Shapley to H.N. Russell, 16 September 1929, Shapley papers, Harvard University Archives.

Horrocks (who probably did not take holy orders) identified an error in Kepler's prediction for the next transit of Venus, and made an informed guess about the right one. The phenomenon, an eclipse of sorts, occurs when Venus comes between the Sun and the Earth, and appears as a tiny black disc passing across the face of the Sun. Horrocks alerted his friend William Crabtree to the rare event, and on 24 November 1639 they were the first Europeans to observe the transit.[53]

Among the observatory staff who entertained the Milnes was the celebrated Annie Cannon,[54] the doyenne of spectral type classification, who knew Oxford well as she habitually spent her summers at the university observatory. She devised a system, largely in use today, to catalogue a quarter of a million stars into forty classes, each one assigned a unique 5-digit reference according to location, brightness and colour, and she could swiftly recall a remarkably large number of them. For her colossal life's work Oxford University awarded her an honorary doctorate in 1925, the first woman to receive one.

Often for dinner the Milnes ate with the Norwegians, Svein Rosseland[55] and his wife, who added an agreeably international dimension. Svein Rosseland could read textbooks in Italian and German, and had held appointments in Norway, Denmark and America. Milne established a good rapport with him, and admired his style of writing in the *Handbuch der Astrophysik*: 'You have done it infinitely better than I would have done. ... I am never clear on a topic until I have gone into details which probably bore the reader.'[56]

The Milnes had a 'topping'[57] final week as guests of the Plasketts at their Brattle Street house. This gave Milne and Plaskett the closest of opportunities to get to know each other and would redound to Oxford's

---

[53]Transits have the peculiar pattern of occurring in pairs eight years apart, every hundred years or so. The last transit was on 6 June 2012, and the next pair will occur on 11 December 2117 and 8 December 2125.

[54]Annie Jump Cannon (1863–1941) published her ten-volume catalogue of stellar spectra in 1924.

[55]Svein Rosseland (1894–1985) created Norway's Institute of Theoretical Astrophysics.

[56]Letter from Milne to S. Rosseland, 18 January 1928, SR, RA/PA-1514.

[57]Letter from Milne to A. J. Cannon, 15 September 1929, Cannon papers, Harvard University Library.

scientific benefit. On 15 September the Milnes sailed from Boston Harbour aboard the White Star liner SS *Cedric*. The Dunhams sent them a farewell telegram of good wishes and Milne replied, 'Our memories of our trip pivot round you two, and the Thompsons and the Plasketts. Somehow you wound magic into it all.'[58] Primed with American optimism, Milne was ready to fight his corner at Oxford.

---

[58]Letter from Milne to T. Dunham, 10 November 1929, EAM, MS.Eng.misc.b.429.

Chapter 9

Cut and Thrust

*I see that a man must either resolve to put out nothing new, or to become
a slave to defend it.*[1]
God offers to every mind its choice between truth and repose.[2]

Milne was Oxford's first mathematical physicist and because he fell outside the existing mould in a highly conservative environment he struggled
to gain recognition for himself and his subject. In his hard-hitting inaugural lecture[3] he made his case forcefully, aware that sitting in the front row
listening to him were the Vice-Chancellor and other dignitaries, some of
whom had questionable sympathy for science. He did not mince his
words.

> It should hardly be necessary even to mention this, but one is more and
> more astonished by the failure of many men of letters to understand what
> the man of science is after. ... Science is given the attributes of a relentless
> machine, crushing the spirit of man in proportion as it ministers to his body.
> ... [Yet] the man of science embarks on a quest of adventure. ... He is of
> the same stuff as anybody else engaged in creative work.

Milne emphasised that his subject was not merely an adjunct to experimentation but a subject in its own right that required discipline and judgement. It could give birth to fresh ideas. Tracking down mathematical
analogies from seemingly disparate situations might reveal a common
basis and identify unity and orderliness — one of Milne's favourite

---

[1] Isaac Newton, quoted by Milne in a letter to *The Observatory* 1949.
[2] Marked by Milne in his copy of *Essays and Representative Men* by R.W. Emerson,
London: Collins, 1850, p. 202.
[3] *The Aims of Mathematical Physics*, Oxford: Clarendon Press, 1929.

themes. The greatest task of a mathematical physicist was to conjecture on the laws of nature; his own mission was to construct stellar models and infer their properties. Explanations of observations would follow, but 'it is not of explanations that the glories of mathematical astrophysics consist.'

Milne's scientific isolation at Oxford did him no favours and he lacked someone with whom he could discuss his stellar ideas and who might persuade him to overhaul his wilder conclusions. But boosted by fresh energy from his American trip, he was ready for battle on all fronts. The words of Curtiss, Dunham, Shapley and Rosseland were ringing in his ears, and they did not think his analysis of the internal structure of a star as preposterous as Eddington did, yet Milne 'did not want to start a storm'[4] until he had shed his last doubts. Despite the mental turmoil of mounting an attack on Eddington's mass/luminosity relationship, he put his ideas to the Royal Astronomical Society in November. The Fellows assembled in a cordial atmosphere, as was their custom, and took their places on the upholstered leather benches in the ground-floor room. Usually a brief courteous discussion followed each talk. After Milne had finished, the programme was running late but, undeterred, Eddington rose to his feet and disdainfully remarked, 'My interest in the rest of the paper is dimmed because it would be absurd to pretend that I think there is the remotest chance of his being right.'[5] While it was commonly known that Eddington could be aloof, this crushing salvo struck many of those present as downright rude and unnecessarily patronising to a younger man. At first Milne brushed off the riposte: 'Eddington said openly it was all nonsense, and we are all good friends',[6] but Eddington's dismissal rankled.

The era of specialisation had not yet arrived and astronomers grasped the main thrust of most papers, so the issue of *The Observatory* that carried the verbatim report of the meeting caused a stir. In Britain, if not in America, Eddington was an acclaimed national figure, revered by astronomers as though he were a minor prophet, and here was someone daring to defy him. Month by month, as the dispute unfolded, astronomers followed its machinations closely.

---

[4] Letter from T. Dunham to M.P. Dunham, 1929, typed copy in the author's possession.
[5] *The Observatory*, 52 (December 1929), p. 349.
[6] Letter from Milne to T. Dunham, 10 November 1929, EAM, MS.Eng.misc.b.429.

Had the mechanisms of stellar energy been understood, the controversy might never have occurred. Ten years would elapse before Hans Bethe,[7] and independently Carl von Weizsäcker,[8] formulated the 'carbon cycle' that explains how carbon acts as a catalyst to trigger the release of energy through a sequence of reactions that converts hydrogen nuclei into a helium nucleus. At this stage the neutron was not in the reckoning either.

Such knowledge as there was, sparse and rudimentary, invited speculation about how stellar evolution relates to the stellar energy that keeps a star shining. Eddington assumed that a star is gaseous throughout, and that its density gradually increases towards its centre. Intuitively he assumed, and assumed correctly as it turned out, that the relationship between mass and luminosity does not depend on energy released in the interior. Yet his model contained a perplexing puzzle: at the centre of the star the calculated temperature was not high enough to sustain thermonuclear fusion, the supposed source of the star's energy. Furthermore there was a substantial discrepancy between his theoretical value of the opacity (its ability to transmit radiation) and the observed value.

Milne never liked tinkering with someone else's theory, and vastly preferred to begin afresh from a new angle. He wished to relate the star's structure to its energy output, and while readily acknowledging an empirical link between mass and luminosity, he assumed they were independent of each other because he was convinced that the surface of the star must play a part in the way it burns its energy. This Eddington denied. Milne began his mathematical analysis from data about the surface of a star and systematically worked inwards to infer conditions at its centre. One of his configurations matched Eddington's, but Milne's mathematics also threw up improbable and surprising alternatives: stars with extremely dense centres at the desired high temperatures, and stars of huge radii. Different cores, he surmised, corresponded to different types of stars.

Given the embryonic level of astrophysics, there was no decisive way of deciding between the rival theories, and some felt that either approach

---

[7] Hans Albrecht Bethe (1906–2005). Born in Germany, he emigrated to Britain in 1933, and in 1935 to America, where he joined the faculty of Cornell University. Nobel Laureate 1976.

[8] Baron Carl Friedrich von Weizsäcker (1912–2007). Theoretical physicist and philosopher. He held appointments at Göttingen, Hamburg and Starnberg.

might be valid.[9] To a large extent the dispute hinged on the choice of assumptions, for, as Milne later admitted, astrophysics suffered from a plethora of assumptions. 'What the stars are really like inside we shall never know in this generation', but we can produce 'a definite set of theorems ... against which future observations can be projected.'[10]

That Milne is more remembered for the controversy than for his pioneering work in stellar atmospheres is a pity. The substance of some thirty papers he wrote in the 1920s do not carry his name, as explained in Chapter 5, whereas the controversy is ablaze with names.

The notoriety of the dispute stemmed as much from the intransigence of the protagonists as from their vivid language. Parodying a well-known examination question, Eddington mocked Milne by comparing his model of a star to an apple pie thirty-four feet high, and declared that Milne was 'between the devil and the deep blue sea — or rather me and the deep blue sea'.[11] For his part, Milne suggested Eddington inhabited 'Flatland', and if he would but make the 'mental effort to scramble up the sides of the trench he would find the surrounding country totally different',[12] and Milne marshalled Humpty-Dumpty into the combat arena.

Although Milne's high regard for Eddington intensified the pain of his persistent onslaught, Milne went out of his way to make generous references to him, hailing him as a 'friend and teacher'.[13] Oversensitive and stubborn, Milne could not help thinking of him, 'waking and even sleeping'.[14] Although Eddington's accusations of 'sophistry and mysticism'[15] cut Milne to the quick, they evoked sympathy from Harlow Shapley and his colleagues who were 'startled at Eddington's temper and what seems to us an evasion of your points'.[16]

---

[9] Letter from A. Wesselink to the author, 18 September 1990.

[10] Letter from Milne to E.C. Stoner, 24 January 1931, Stoner papers, Leeds University.

[11] *The Observatory* 54 (February 1931), pp. 34, 36.

[12] *The Observatory* 53 (November 1930), pp. 305–308.

[13] *Nature* 125 (22 March 1930), p. 454; (10 May 1930), p. 708.

[14] Letter from Milne to S. Rosseland, 25 January 1930, SR, RA/PA-1514.

[15] Letter from Milne to W.G. Thompson, 22 November 1930; a copy is in the author's possession.

[16] Letter from H. Shapley to Milne, 7 April 1930, Records of Harlow Shapley 1921–1956, Harvard College Observatory, UAV 630.22, Box 42.

As the contretemps snowballed in complexity, Milne and Eddington hammered away in *Nature* and *The Observatory*, and in face-to-face skirmishes at RAS meetings. Some claimed that Eddington won on paper but lost in debate because he lacked spontaneity and left his sentences unfinished. Milne, on the other hand, was at his best on his feet. He relished the cut and thrust of debate, his words gushing forth in a torrent. Their verbal jousting became a spectator sport for even the purest of mathematicians, G.H. Hardy, who stated that he did not care two straws what the stars were really like but had a mild preference for Milne's theory because it took account of all solutions.[17] This was ironic, for at the root of the controversy lay Milne's extravagant, uncritical interpretations of his solutions without regard for the physics. That Milne gave prominence to the solutions which suited his argument brought forth justified accusations of irrationality. But, shamelessly hitching himself to Hardy's mathematical bandwagon, Milne declared that he was not concerned with whether the stars were actually constructed as he proposed. 'We can only find out by building [a model of] a star and seeing how it works.'[18]

Milne's assault stimulated widespread interest. Better theory arises from exposing false premises, inconsistencies and illogical conclusions, and progress may lie along an unchartered path strewn with hidden pitfalls, stumbling blocks and tempting blind alleys. Luck, personal interaction and body language may play their part.

Sir Oliver Lodge expressed surprise at the quantity of argument, considering that Eddington and Milne agreed on many major points. Approval, in varying measures, came from Milne's colleagues in Germany, Holland and America. Lindemann weighed in for him, as initially did Sir James Jeans, who also believed in the independence of mass and luminosity. The venerable Irish physicist Sir Joseph Larmor[19] 'saw what [Milne] was driving at',[20] and Sir William Bragg asked Milne to give two talks to students at the Royal Institution.

---

[17] *The Observatory* 54 (February 1931), p. 40.

[18] Letter from Milne to T. Dunham, 25 December 1929, EAM, MS.Eng.misc.b.429.

[19] Sir Joseph Larmor (1857–1942). Professor of Mathematics at Cambridge from 1903 to 1932. Member of Parliament for Cambridge University from 1911 to 1922.

[20] Letter from Milne to W.G. Thompson, 30 March 1930; a copy is in the author's possession.

In the summer Eddington invited Milne to a private talk, although he 'feared it would be rather like an irresistible force meeting an immovable body — but he wouldn't say which was which'.[21] They had a 'dramatic' discussion lasting four and a half hours with a break for a 'friendly lunch' during which 'neither of us budged an inch'.[22] The matter remained unresolved.

If previous accounts of the quarrel portray it as intensely adversarial, this is not entirely accurate. According to an eyewitness, Milne and Eddington conducted themselves with reasonable politeness,[23] and their clashes were neither as acrimonious nor as vehement as the earlier slanging matches between Jeans and Eddington. Milne and Eddington did not trade insults so much as pick at the labyrinth of entangled assumptions and postulates. Their exchanges at RAS meetings resembled family squabbles and at the dining club afterwards they were quite likely to sit at the same table. Admittedly there was disagreement, but it was not disagreeable. J.B.S. Haldane, who knew Milne before and after his *encephalitis lethargica,* thought the illness made him more polemical.[24] Had the astronomical community been larger and more diffuse, the spat would almost certainly have been less personal and less polarised.

Leon Mestel assesses the controversy: Eddington's intuitive approach to the structure of a homogeneous, 'main sequence' star like the Sun turned out to be basically valid. If the inward gravitational pull is balanced by the pressure exerted by a perfect classical gas, the required temperatures are so high that the stellar material will indeed behave like a perfect gas. The spontaneous flow of radiation from the hot interior to the surface is balanced by sub-atomic energy liberation ε per gram. For his 'standard model', Eddington adopted a special — and quite unjustifiable — approximation, equivalent to making ε depend only weakly on the temperature. This yielded a luminosity L dependent strongly on the mass M but independent of the radius R, yet remarkably close to the

---

[21] Letter from Milne to T. Dunham, 11 July 1930, EAM, MS.Eng.misc.b.429.

[22] Letter from Milne to E. Rutherford, 1 August 1930, SHQP; and letter from Milne to G. Milne, 5 August 1930, EAM, MS.Eng.misc.b.423.

[23] Evans, D.S. *The Eddington Enigma*, Princeton: Xlibris, 1998, p. 133.

[24] Letter from G.J. Whitrow to N. Hetherington, 10 July 1978, in the author's possession.

observed values. Elsewhere in his *Internal Constitution of the Stars* Eddington conceded that ε would in fact depend strongly on temperature, as was subsequently shown to be the case by Bethe, von Weizsäcker and others. Later, Cowling, Hoyle,[25] Lyttleton[26] and Schwarzschild showed that modifications of Eddington's treatment yielded L fixed essentially by radiative transfer and depending primarily on M, while R is fixed by the temperature-sensitive, energy balance condition. With hindsight, one can say that the temperature-sensitivity of ε is essential for Eddington's approach.

Milne's permanent contribution to stellar astrophysics is in stellar atmospheres, where he shares the honours with Eddington. In the simplest case, the Milne–Eddington integral equation provides mathematical conditions for the escape of radiation. Eddington argued that the stellar surface is 'passive', adjusting its temperature so that it radiates the luminosity supplied from below. Milne, however, argued that there were solutions alternative to Eddington's, with dense, so-called 'degenerate' cores, and with a structure that depended on the surface conditions. Investigations by Cowling, Chandrasekhar,[27] Russell and Strömgren[28] failed to support him. However, because of the temperature-sensitivity of the hydrogen fusion process, the homogeneous, main sequence star evolves into a red giant star, with a dense, burnt-out helium core, surrounded by an energy-liberating shell. In their pioneering study, Hoyle and Schwarzschild found that the stellar envelope surrounding the core must 'boil' — i.e. it becomes largely convective. Their theory predicted an evolutionary path in the stellar luminosity–colour diagram in reasonable agreement with observation. So, as noted by Fred Hoyle in his autobiography *Home is Where the Wind*

---

[25] Sir Fred Hoyle (1915–2001). Astronomer, broadcaster and author. Plumian Professor at Cambridge from 1958 to 1972. He was an outspoken critic of 'big bang' theories of the Universe.

[26] Raymond Arthur Lyttleton (1911–1995). Astrophysicist and author.

[27] Subrahmanyan Chandrasekhar (1910–1995). Astrophysicist. In 1929 he left India for England. In 1936 he emigrated to the USA, holding appointments at Yerkes Observatory, Wisconsin and the University of Chicago.

[28] Bengt Georg Daniel Strömgren (1908–1987). In 1940 he succeeded his father as Director of the Copenhagen Observatory. Director of the Yerkes and McDonald Observatories from 1957 to 1967.

Blows, Milne's claim that the Milne–Eddington surface condition may react back on the gross stellar structure does turn out to be pertinent, but ironically, for inhomogeneous, evolved stars, very different from the homogeneous, non-convective models that he and Eddington studied. Likewise, Hayashi[29] found that the surface boundary condition plays an active role in fixing the path of the star.[30]

Eventually the controversy petered out and resulted in a stalemate. Milne resented frittering away his energy in defending his ideas and told Rutherford that in the future he would apply his mind constructively. Pursuing his stellar model of a dense core inside a gaseous envelope — his simile of the yolk of an egg caught the eye of the press — he investigated the energy changes that make a star shine. In Bristol at the 1930 meeting of the British Association for the Advancement of Science he astonished his listeners with the sweeping announcement that *every* star, once it has exhausted its fuel, would erupt in a cataclysmic explosion of stupendous brightness, a nova. The star would blow off its outer shell and form a binary pair, with the original star collapsing into a white dwarf, a small dying star with no apparent source of energy. While this is not the case for all stars, H.N. Russell thought the proposal 'very pretty'.[31] 'Fowler has lectured on it at Copenhagen and Bohr accepts it … [Eddington] finally admitted grudgingly that it could not be dismissed offhand.'[32] Yet still Milne felt under siege: 'The wolves are out on the trail, and the big stick of authority is to be used on my work.'[33] 'Scientific ostracism is threatened an unbeliever.'[34]

The dispute with Eddington did not monopolise Milne's thoughts. In the course of exploring the mechanisms that cause Sunspots to circulate,

---

[29] Chushiro Hayashi (1920–2010). Japanese astrophysicist.

[30] My thanks go to Leon Mestel for his email, 10 March 2010.

[31] Letter from H.N. Russell to Milne, 27 May 1931, papers of Henry Norris Russell, Manuscripts Division, Department of Rare Books and Special Collections, Princeton University Library.

[32] Letter from Milne to G. Milne, 23 November 1930, EAM, MS.Eng.misc.b.423.

[33] Letter from Milne to H. Dingle, 14 December 1930, EAM, MS.Eng.misc.b.429.

[34] Letter from Milne to W.G. Thompson, 22 November 1930; a copy is in the author's possession.

he found a use for an equation that had lain dormant for a century.[35] At the Solar Physics Observatory in Cambridge he instigated a programme that he hoped would explain the connection between Sunspots and terrestrial magnetic storms, which were a nuisance to radio transmissions. In order to present a German paper on radiative transfer to the RAS he translated it into English.

White dwarf stars now dominated Milne's attention because of their insight into novae and into the processes of stellar evolution. With the telescopes of the day, white dwarfs, about the same size as the Earth, were almost invisible, faint and tiny. Only a handful had been documented, the best known being Sirius B, companion of the bright dog star, Sirius A. Milne suspected that white dwarfs must be plentiful, because novae were common events, but he did not live to see his forecast confirmed. At his death only twenty had been identified. Today a rough estimate is several billion and we now know that white dwarfs occur in a variety of configurations.

These small and peculiar dying stars are unimaginably dense, 100,000 times heavier than water, utterly unlike any Earthly material. R.H. Fowler showed that they are made of 'degenerate' material, electrons freed from their atoms and tightly compressed together. From this starting point Milne investigated their physics and entered his findings for Oxford's Johnson Memorial Prize.[36] Given the university's glorification of prizes, it would do him no harm to win one. The rules stipulated that the candidate place his name inside a sealed envelope bearing a motto, the same motto to label his script. Milne chose quotations from Horace and Lucretius that refer to the paradox of white dwarfs: they are scorched by hot flames yet stiff with cold.[37] Margot plotted graphs, listed tables of data, and typed most of his manuscript, his own typing being atrocious and riddled with errors. In a rush to meet the deadline Milne submitted the last pages in handwriting, then worried lest he lose his only copy, and asked for his script back after adjudication. He secured the prize of a gold

---

[35] Noted by H.H. Turner in *The Observatory* 53 (April 1930), p. 103.
[36] Manuel John Johnson (1805–1859) won an RAS Gold Medal for cataloguing stars in the southern hemisphere before running Oxford's Radcliffe Observatory for twenty years.
[37] 'Hot' refers to temperature, 'cold' to inert degenerate matter.

medal and £65, and adapted his returned script for his Halley lecture in 1932.

Milne knew that he was handicapped by insufficient knowledge about the way degenerate matter absorbs radiation, its opacity, and he failed to recognise Chandrasekhar's prediction of a 'limiting mass', which differentiates the death throes of stars according to size. The smaller ones collapse into white dwarfs, the larger ones change to neutron stars or black holes. For this groundbreaking work[38] introducing the Chandrasekhar Limit, Chandra was to receive a Nobel Prize in 1983.

Chandra's entry into astrophysics lifted Milne's spirits. Chandra was exceptionally bright and at the age of nineteen he left India to study in Cambridge. There, unaffiliated to any college, lonely and unhappy, he felt a social outcast. When Milne, unannounced, knocked on the door of his 'digs' at 1 Park Street and extended a warm hand of friendship, Chandra was both surprised and gratified. He cited Milne as the first person in England to give him encouragement and based his doctoral thesis on an idea suggested by Milne. Their minds meshed. For his part, Milne wished 'other people understood my analysis as completely as you do'.[39] Their abiding friendship survived unblemished despite the scientific differences that cropped up between them. Chandra adopted a courteous policy of only referring to Milne if they were in agreement, and Milne never argued with Chandra in front of other people (Fig. 12).

Chandra admired the crisp precision of Milne's mathematical papers and welcomed Milne's advice in polishing his own for publication. 'The Secretary of the RAS will ask me, "What has this to do with astronomy?" The answer is a great deal, but to establish this you should give some numerical examples'. Again, 'I just want you to give an astronomical flavour to scotch the qualms of the Secretaries.'[40]

---

[38]Edmund Clifton Stoner (1899–1968) and Lev Landau (1908–1968), Nobel Laureate 1962, anticipated this result. See Nauenberg, M. 'Edmund C. Stoner and the Discovery of the Maximum Mass of White Dwarfs' in *Journal for the History of Astronomy* 39 (August 2008), pp. 297–312.

[39]Letter from Milne to S. Chandrasekhar, 29 January 1931, EAM, MS.Eng.misc.b.427.

[40]Letters from Milne to S. Chandrasekhar, 16 and 29 January 1931, EAM, MS.Eng. misc.b.427.

Although Milne was confident that Chandra would be able to explain the opacity of a degenerate gas, he let him know that Bertha Swirles, his former Manchester colleague, was on the same trail. To Milne's consternation their papers arrived on his doormat by the same post, and they were identical, Chandra's in finished form, Bertha's in notes. Milne wanted to make sure that both got published and knew he must act with propriety. He asked Paul Dirac to endorse Chandra's for the Royal Society[41] — ironically the paper came back to Milne for refereeing — which left his conscience clear to forward Bertha Swirles's to the RAS.

This period was outstandingly exciting for physics and astrophysics. In 1929 data from Mount Wilson suggested that the Universe is expanding, a concept first mooted by the Russian cosmologist Alexander Friedmann,[42] then the Belgian Georges Lemaître,[43] and nearly discovered by the American Vesto Slipher.[44] At Mount Wilson, Edwin Hubble[45] and his assistant Milton Humason,[46] who seldom gets fair credit, patiently exposed photographic plates of faint distant nebulae. By making meticulous comparisons they noticed that the images seemed to indicate that the nebulae are receding. The further away they are the faster they travel (Hubble's Law).

The following year, in 1930, after a long, painstaking search, an astronomer at the Lowell Observatory in Arizona found 'Planet X', soon named Pluto, and recently reclassified as a dwarf planet. Then in 1932 two great events at the Cavendish Laboratory resulted in three Nobel Prizes. James

---

[41] Letter from Milne to Chandrasekhar, 7 May 1931, EAM, MS.Eng.misc.b.427.

[42] Alexander Aleksandrovich Friedmann (1888–1925) spent almost his entire life in St Petersburg.

[43] Georges Edouard Lemaître (1894–1966). After graduate studies at Cambridge and the Harvard Observatory he was appointed professor at the University of Louvain in 1925. His theory starts with a 'primeval atom'.

[44] Vesto Melvin Slipher (1875–1969). Director of Lowell Observatory from 1926 to 1952.

[45] Edwin Powell Hubble (1889–1953). As a Rhodes Scholar he studied law at Oxford. He joined the staff of Mount Wilson Observatory in 1919.

[46] Milton Lasell Humason (1891–1972) left school at the age of fourteen and drove up Mount Wilson the mules that carried the materials to build the observatory. He progressed to janitor, night assistant and finally, in 1919, to a staff appointment.

Chadwick[47] identified the neutron, and John Cockcroft[48] and Ernest Walton[49] sensationally 'split the atom', artificially transmuting one atomic nucleus into another by bombarding a lithium target with protons to produce a helium nucleus.

These fundamental discoveries bore strongly on astrophysics, which is concerned with the behaviour of atomic particles in a stellar context — and they sharpened interest. When Niels Bohr spoke at an open meeting of the Oxford University Mathematical and Physical Society held at Somerville College, two hundred people turned up to hear him. Milne was on friendly terms with him and his younger brother Harald,[50] a mathematician, who was the better known of the two in their own country because he had played football for Denmark in the 1908 Olympics. Their faces were remarkably similar, almost like twins, but Milne thought Niels's showed the greater sensitivity.

Niels Bohr and probably Harald spoke at Milne's colloquia to his students. Although he got his room at Wadham he was disgruntled by the university's churlish behaviour towards his guests from overseas. For speakers he drew on his wide circle of distinguished European astronomers: the Director of the Leiden Observatory, Willem de Sitter,[51] who with Einstein in 1932 foreshadowed 'dark matter'; the Dutch astronomer, Anton Pannekoek;[52] the Norwegian Carl Störmer,[53] whose interest in aurorae prompted him to classify them according to shape; and Svein Rosseland. The Milnes willingly had these people to stay in their flat at

---

[47] Sir James Chadwick (1891–1974). Master of Gonville and Caius College, Cambridge, from 1948 to 1958. Nobel Laureate 1935.

[48] Sir John Douglas Cockcroft (1897–1967). Director of the Atomic Energy Research Establishment at Harwell. Master of Churchill College, Cambridge. Nobel Laureate 1951.

[49] Ernest Thomas Sinton Walton (1903–1995). Professor at Trinity College, Dublin. Nobel Laureate 1951.

[50] Harald August Bohr (1887–1951). Professor at Copenhagen Polytechnic and Copenhagen University.

[51] Willem de Sitter (1872–1934). Director of the Leiden Observatory from 1919 to 1934.

[52] Anton Pannekoek (1873–1960). Communist. Author of *A History of Astronomy*, London: George Allen and Unwin Ltd, 1961. Republished by Dover in 1989.

[53] Fredrik Carl Mülertz Störmer (1874–1957). Professor at Oslo from 1903 to 1946. Oxford University gave him an honorary degree in 1947.

104 Banbury Road but the university's attitude towards outsiders was insultingly lukewarm. Since Milne had no direct access to funds for defraying his visitors' travelling expenses, he asked the university for a modest contribution, typically £5 or £10, and was disgusted to be shuffled from one body to another. In a 'rumpus with the University authorities ... [I] had to complain rather strongly. Oxford is in fact strong as a collection of colleges doing teaching work, but very poor as a university — no cohesion, little appreciation of scholarship or research, low standards, attachment to examination rather than research standards. I foresee that I shall come into conflict with the University in many ways.'[54]

His words were prescient. The sale of the Radcliffe Observatory embroiled him in another wrangle, a complicated debacle[55] that wrenched his loyalty between astronomy and the university. The background to the sorry saga was the entwined history of Oxford's two observatories. Until the latter part of the eighteenth century studying the stars was mainly the domain of wealthy individuals. Then, in 1772, the Radcliffe trustees agreed to the request by the university's Savilian Professor to create an observatory from Dr John Radcliffe's benefaction. A century later, largely through the university's carelessness, the Savilian Chair lost its tie with the Radcliffe Observatory, and the university built its own observatory in the Parks less than a mile away. In 1902 the Radcliffe Observatory was modernised and Milne relied on its superior resources. He was on close personal terms with Harold Knox-Shaw,[56] its director, called the 'Radcliffe Observer', who obliged Milne by stocking the library with periodicals he wanted.

Although Milne had little connection with the OUO, he had long known its genial director, Henry Turner, an astronomer of international repute and sometime president of the RAS dining club, who could be relied on for a song or witty speech. During the past forty years Turner had

---

[54]Letter from Milne to G. Milne, 23 March 1930, EAM, MS.Eng.misc.b.423.

[55]I am indebted to Roger Hutchins for steering me through the labyrinth of information that he analysed in his *British University Observatories 1762–1939*, Burlington: Ashgate, 2008, pp. 319– 370.

[56]Harold Knox-Shaw (1885–1970). Director of the Radcliffe Observatory in Oxford from 1924, and in Pretoria, South Africa, from 1939.

immersed himself in two gargantuan tasks: the astrographic catalogue, which ambitiously aimed to photograph the entire sky, and compiling data on earthquakes. Tied up with these commitments, he neglected to provide practical instruction to students, and he was unreceptive to astrophysics. Turner locked the OUO in a time warp. Its equipment was obsolete and the university was unlikely to replace it as long as Turner was in charge. They were at loggerheads because twice the university had refused to grant him an official residence near the OUO, whereas every other major observatory had nearby accommodation for its director.

Plans to sell the Radcliffe Observatory were well advanced. Sir William Morris (Lord Nuffield)[57] had generously offered to pay £100,000 for its nine-acre site to enable the adjoining Radcliffe Infirmary to expand its postgraduate facilities. With the proceeds the Radcliffe trustees intended to construct a new observatory in South Africa, where the clearer skies made for better observing conditions. In the summer of 1929 a party, which included Knox-Shaw (and his brother Tom,[58] a mathematician friend of Milne's), chose a hilltop near Pretoria as a suitable site. But to build outside the United Kingdom the trustees needed the consent of the charity commissioners — and this was the Achilles heel. If the university contrived to block this consent, the Radcliffe windfall might yet remain within the vicinity of Oxford and regenerate university studies of astronomy. Legal argument seemed thin, but grounds for objection could be cobbled together by drawing on historical precedent.

Knox-Shaw and Turner envisaged a mutually profitable alliance, sharing photographic data and a travelling studentship. (Several American and German observatories benefited from outstations in the southern hemisphere.) Confident in Knox-Shaw and out of loyalty to Turner, Milne backed the move for its obvious advantages to astronomy. The co-operative scheme needed approval from the university's Hebdomadal Council, but, in a show of independence, the Radcliffe trustees refused to disclose their detailed plans. The university was indignant at this

---

[57] William Richard Morris, Viscount Nuffield (1877–1963). Industrialist and philanthropist.
[58] Thomas Knox-Shaw (1886–1972). Master of Sidney Sussex College, Cambridge, from 1945 to 1957.

perceived discourtesy, particularly as it had liaised harmoniously over the recent incorporation of the Radcliffe Science Library into the Bodleian Library.

Lindemann, ever eager to upgrade the university's science facilities, whipped up opposition to the move to South Africa. He argued that there was a moral duty to prevent the Radcliffe endowment being siphoned off beyond the reach of the university. Exploiting the university's pique and denigrating Turner, he bludgeoned the university's Board of Physical Sciences into agreeing that the loss of the Radcliffe Observatory would damage studies in physical sciences. Milne, a member of the board, deplored Lindemann's sledgehammer tactics, and stood up to his bullying and cast a dissenting vote. Lindemann brazenly and publicly proceeded to step up his campaign. With his ally and subsequent biographer Lord Birkenhead,[59] the high steward of the university, he orchestrated a rude attack on Turner in *The Times*.[60] Birkenhead's intemperate accusations incensed Milne, who pencilled 'nonsense', 'irrelevant' and 'non-sequitur' in his copy of the paper. Feelings ran high.

Milne was barely into his second year at Oxford. Peeved by the university's indifferent welcome and only partially assimilated into its milieu, he did not take an 'Oxford' viewpoint. His sympathies lay with astronomy. Milne sat on the British National Committee for Astronomy, which overwhelmingly backed the Radcliffe trustees and made its view clear by publishing a letter[61] to *The Times*, listing the names of its seventeen members, the cream of Britain's astronomers.

Wanting to effect reconciliation and find a compromise, Milne tried to diffuse the sour wrangling that threatened to split Oxford's scientific community. Beyond beseeching Lindemann to tone down his belligerence, Milne lobbied, mediated, organised meetings, composed reports and wrote letters in all directions. He pressed the Vice-Chancellor, F.W. Pember,[62] to dissuade members of the Hebdomadal Council from

---

[59] Frederick Winston Furneaux Smith, Earl of Birkenhead (1907–1975). Historian.

[60] *The Times* on 28 March; 17, 23, 26, 29 April and 2, 3, 10, 15 May 1930.

[61] *The Times* (17 May 1930) and *Nature* 125 (24 May 1930), p. 776.

[62] Francis William Pember (1862–1954). Lawyer. Warden of All Souls College from 1914 to 1932. Vice-Chancellor from 1926 to 1929.

grumbling about the trustees. Joining forces with Knox-Shaw, Milne urged Sir Frank Dyson,[63] the only astronomer among the Radcliffe trustees, to persuade the trustees to foster more friendly relations with the university, which pleaded that were the trustees to be less cagey, the university would dampen its opposition. But the trustees remained implacable. They resented Lindemann's meddlesome interference. Attitudes hardened and hostilities deepened.

By June 1930 Milne had changed his allegiance. He now supported the university, having taken account of the university's duty to educate students. His swing in allegiance came hard on the heels of a visit to Oxford from his new friend Harry Plaskett, who stressed that it was vital for graduate students to have access to an observatory. Keen to attract and teach students, Milne felt that the Radcliffe money should go to improving facilities, knowing that under Turner's regime they had not. Furthermore, the university had never had a large endowment for science, and the chance to channel the Radcliffe nest egg to the university's advantage was not to be let slip. To make a strong case required concrete proposals. Milne favoured an Astrophysical Institute, along the lines of the one Svein Rosseland was setting up in Oslo. If it embraced astronomy, theoretical astrophysics, geophysics and meteorology, Oxford would be pre-eminent in cosmic science. An alternative proposal, which Milne also endorsed, was to build a new observatory on Headington Hill.

In August the sudden death of Turner irrevocably altered the university's position. Overnight the fortunes of the OUO looked up. If Turner's successor exerted drive and enthusiasm he could revive studies in astronomy. Meanwhile, before making the crucial appointment, a committee that looked into the situation concluded that the university's paramount need was for a modern, well-equipped observatory. In a supporting affidavit Milne pointed out that because Oxford was unable to provide him with stellar data he had to resort to Cambridge, Greenwich, Leiden and Mount Wilson. His argument that the OUO's current primitive facilities discouraged studies in astronomy decisively swayed the university's registrar,

---

[63] Sir Frank Watson Dyson (1868–1939). Astronomer Royal for England from 1910 to 1933.

Douglas Veale,[64] to make a legal challenge to the trustees. It was not until 1934 that the case in Chancery was heard, and the university lost. From the ashes, the university salvaged a link with the Radcliffe Observatory in the form of the Radcliffe Travelling Fellowship,[65] whereby an astronomer divided his time between Oxford and South Africa.

The appointment of Turner's successor was of critical importance. Milne was in no doubt that the best man for the job, who possessed exactly the right personality and experience, was Harry Plaskett. Milne influenced the appointment at every stage. First, though, he had to persuade Plaskett to apply for the chair, which carried a lower salary than he earned at Harvard. Plaskett was under no delusion about the 'general Oxford attitude that a man is allowed but not encouraged ... to do research', but 'to be with Milne [was] worth taking a chance'.[66] Milne put his name forward and nursed his application. By praising the 'provocative'[67] quality of Plaskett's work, Milne tellingly revealed what he himself most valued. Plaskett was chosen and Milne was elated. Although a newcomer to Oxford, he had secured his preferred candidate. Moreover, he had gained a sorely needed colleague after his own heart.

Milne's personal knowledge of Plaskett carried weight and he did not come for interview, but before taking up the Savilian Chair in June 1932 he travelled to Oxford at his own expense to discuss the future of the OUO. The Milnes were delighted to have him to stay, and yet again Milne was annoyed at the university's parsimony in refusing to make a contribution to Plaskett's transatlantic fare. This was not the way to treat an incoming professor.

In a clever stroke of diplomacy and shrewd judgement, Plaskett established both his autonomy and his authority by putting forward a pragmatic plan for the OUO, which was entirely independent of the Radcliffe case. He proposed specialising in solar physics. Studying the Sun, our closest

---

[64] Sir Douglas Veale (1891–1973). Administrator. Registrar of Oxford University from 1930 to 1958.

[65] The first holder was the Dutch astronomer Herman Zanstra (1894–1972).

[66] Letter from H.H. Plaskett to T. Dunham, 13 December 1931, Dunham papers, Niels Bohr Library and Archives, American Institute of Physics, College Park, MD USA.

[67] Letter from Milne to F.A. Lindemann, 5 October 1931, Cherwell papers, Nuffield College, Oxford.

star, suited Oxford's grey weather, and buying solar equipment would be less expensive than replacing the old telescopes.

At about the same time that Milne was entangled in the Radcliffe fracas, he was caught up in an unfortunate tiff with Tom Cowling. Feeling alone in London, Cowling wrote lengthy weekly letters to Milne. Milne replied, but after a bit asked Cowling to write less frequently. He complied but grew steadily more critical of Milne's work, until Milne's patience snapped and in a stinging letter he accused Cowling of disloyalty. Cowling was desolate. He had no inkling that he had overstepped the mark, nor of the emotional strain that Milne was under from many quarters. After Sydney Chapman interceded they patched up their differences, but the breach left scars.

A happy respite from these travails was Einstein's visit to Oxford in May 1931. Lured by the promise of an honorary degree, he agreed to deliver the Rhodes Lectures. On three consecutive Saturdays he lectured in German (a translation was available). Milne got to know him — they spoke in French — at a number of lunches and dinners, at Wadham and elsewhere,[68] and Milne was pleased that Einstein came to his colloquium to hear him talk about novae.

By the end of June Milne felt exhausted. He had been to Manchester and Edinburgh as an external examiner, and was inundated with papers for review. Ahead lay the preparation of a lecture in French for the Sorbonne, one for Berlin in July at the request of Erwin Schrödinger,[69] the creator of wave mechanics, and a three-day course at the Royal College of Science. Foolishly, he did not take a break, and by the autumn he was dog-tired.

Usually he enjoyed the British Association for the Advancement of Science meetings but he arrived jaded for its centenary. Under the presidency of General Jan Smuts,[70] former Prime Minister of South Africa, two thousand people converged on London for the extensive programme of

---

[68] Dixon, Hardy, Love and Milne attended a dinner for Einstein at Merton College. Wilson, R.J. (2007). 'G.H. Hardy's Oxford Years', courtesy of Robin J. Wilson.

[69] Erwin Schrödinger (1887–1961). Austrian physicist. He held appointments at Stuttgart, Breslau, Zurich and Berlin, and was in Oxford from 1933 to 1936. Later, after 17 years in Dublin, he returned to Austria. Nobel Laureate 1933.

[70] General Jan Christiaan Smuts (1870–1950). Statesman, soldier, philosopher. Prime Minister of South Africa from 1919 to 1924 and from 1939 to 1948.

talks, exhibits and expeditions. The highlight was a discussion titled, 'The Evolution of the Universe', which attracted such a large audience that at the last minute the debate was relocated from South Kensington to the Central Hall, Westminster. (When Milne lectured on the same theme in Glasgow fifty people were turned away.)

On the platform Milne sat alongside the illustrious panel of scientists and theologians: Bishop E.W. Barnes, Eddington, Jeans, the plump and cheery Abbé Georges Lemaître, Sir Oliver Lodge, the American Robert Millikan,[71] Willem de Sitter and General Smuts. Jeans reassured his listeners that the Earth was in no imminent danger of coming to an end, but when the Universe did eventually run out of energy it would reach a state of inert stagnation, a so-called 'heat death'. Other speakers endorsed this prediction on the grounds that the 'heat death' was an inevitable sequitur of the second law of thermodynamics, which dictates ever-increasing disorder. The press dramatised this scenario with headlines such as, 'Scientists Probe Riddle of the Universe' and 'Is the Universe Merely an Ash-Heap?'[72]

Milne contested the dogma of this gloomy prediction and intended to say so. The second law is valid only if a system is closed, and Milne did not think the Universe met this condition owing to its expansion. He was not alone in disputing it. Sir Joseph Larmor 'did not hold a brief for entropy',[73] neither did Sir Oliver Lodge nor Robert Millikan[74] nor several other notable people.[75] Nevertheless the 'heat death' verdict was well entrenched. Milne had suffered the humiliation of the Royal Society

---

[71] Robert Andrews Millikan (1868–1953) measured the charge on the electron. President of the California Institute of Technology from 1921 to 1945. Nobel Laureate 1923.

[72] *The Scotsman* and *The Daily Mail*, 30 September 1931.

[73] Letter from J. Larmor to Milne, 21 June 1930, in the author's possession.

[74] Milne, E.A. (1952). *Modern Cosmology and the Christian Idea of God,* Oxford: Clarendon Press, p. 146.

[75] For example, Walther Nernst (see Kragh, H. 'Walther Nernst: Grandfather of Dark Energy?' in *Astronomy and Geophysics* 53 (February 2012), p. 124), the mystic-explorer-diplomat Sir Francis Younghusband (see his 1933 work *The Living Universe*, London: Murray, pp. 1–4, 14) and J.S. Haldane (see his 1932 work *Materialism*, London: Hodder and Stoughton, p. 116).

rejecting a paper[76] from him on the subject, and he knew his position on it was regarded as an affront to orthodoxy. Before the debate he apprised Jeans of his intention. Jeans frowned. He had thoroughly disagreed with Milne's heretical paper and intimated to Milne that he should not take this line at the debate. Probably today Milne would not be dismissed so summarily. There is no consensus of opinion about our fate and we do not understand the part played by the mysterious dark energy. The current menu offers the choice of a big crunch of annihilation, a perpetual cycle of death and rebirth, a ripping apart, or a situation in which the Universe accelerates forever and enjoys immortality.

While submitting to Jeans's authority, Milne was not averse to stirring up a little mischief. When it was his turn to speak, he explained how novae provide evidence for an evolving Universe. Casting a glance at Bishop Barnes, who smiled indulgently, Milne cited a Biblical simile: a collapsing star burning off its outer layers was like Samson losing his strength after his locks were shorn. Milne concluded by affirming that God the Creator underpinned the scientific world, a provocative remark that was bound to aggravate some scientists. Milne was courting trouble.

---

[76]The Royal Society received the manuscript on 10 June 1930 and returned it with referees' comments on 2 July 1930; it is now in the author's possession.

Chapter 10

Family versus College

Married Fellows were treated as honorary bachelors.[1]

A month later my parents moved into a three-storey house not far from their flat, which they had outgrown with the birth of my sister in 1930. Typical of north Oxford's Edwardian Gothic architecture, 19 Northmoor Road was in a network of streets populated by dons with families. My father quipped to Svein Rosseland that they had resisted the temptation of calling their daughter Stella or Astra, and chose Eleanor. I arrived in 1933 and when the RAS dining club toasted my health, my father replied that he lived in an expanding Universe (Fig. 13). My brother Alan followed in 1938.

My parents enjoyed a comfortable but not lavish lifestyle. Entertaining and organising came naturally to my mother and she kept meticulous records: a dinner book of guests and what they ate; baby books of our diet, physical development, vocabulary and even dates of our playmates' birthdays. Like her Fiddes aunt she was good with servants and when the household was in full swing we had two live-in maids, Violet and Gladys, a cheerful stout charlady Mrs Reid, a daily cook and a weekly seamstress. Given the lack of ready-made clothes and domestic appliances, this was not exceptional for a professional family.

My father made a point of being accessible to students and they would drop by on the off chance for a chat. In his ground-floor study they would sit in easy chairs around the fireplace under engraved prints of Galileo, Newton and J.J. Thomson, who gazed down as though scrutinising the proceedings. Opposite, on shelves that stretched from floor to ceiling, were books and beloved runs of periodicals that my father maintained fastidiously.

---

[1] Harrison, B. (ed.) *The History of the University of Oxford* vol. 8, Oxford: Clarendon Press, 1994, p. 86.

Next to my father's study, the drawing room led through French windows to the garden, a lawn surrounded by herbaceous beds and espaliered fruit trees along brick walls. At the back of the house, beyond the scullery, larder, pantry and cosy kitchen with the anthracite boiler, was the all-important back door. A rap on it announced the baker, his basket over his arm, his docile horse and cart left unattended in the road, or a delivery boy on his bicycle with meat, fish, groceries and vegetables. Sundry other purchases came by van, for to be seen carrying a parcel was considered undignified.

Staid conventions constrained social behaviour, and Oxford tended to be dressy — often a dinner jacket in the evening or tails if the Vice-Chancellor were present. When H.N. Russell gave the Halley Lecture, Plaskett advised him to pack both sets of evening clothes. It was unthinkable for my mother to leave the house without a hat and imperative for her to wear the correct gloves. As a university wife she was expected to 'pay calls', and, having endured these tiresome afternoons in Manchester, she knew the protocol of visiting cards and the pitfalls of straying from etiquette and causing unwitting offence. As to topics of conversation, any hint of politics, religion or money was taboo.

With ample domestic assistance, my mother had time to help my father, who liked her to participate in his endeavours as much as possible — he made sure she had a ticket to hear Einstein lecture. He wanted her to share his veneration for his scientific heroes and read aloud to her from the writings of Larmor (whose complete works he proudly owned), Kelvin, Poincaré, Rayleigh and Stokes.

My father was at his best with us in our garden (Fig. 14), on walks in Port Meadow pointing out wildflowers, and when blackberrying at Eynsham, where he hooked the highest branches down with his walking stick for us to pick. He passed on his knowledge of heraldry but, ever the perfectionist, expected our drawings to be faultless and fussed if our colouring went beyond the outline. We had glorious summer holidays at our grandfather's rambling house in Dornoch, joined by Jean's family and relays of Fiddeses. We thought the world of our playful grandfather, who wore the kilt and in Scots fashion dipped each spoonful of his salted breakfast porridge into a cup of milk before slurping it through his moustache. For all his brains, he had no business sense, and was hopelessly soft-hearted. While practising as an advocate in Aberdeen he accepted a large halibut in lieu of a fee from an impecunious fisherman, not realising that the fish would rot before he could finish eating it.

His amiable inefficiency and dubious financial schemes that invariably ended in disaster did not endear him to my father, who addressed him as Mr Campbell. Nonetheless, my father entered into the spirit of the holiday, building us sandcastles, picnicking, attending the Highland Games and venturing on family jaunts to Bonar Bridge and John o'Groats.

Although we spent our summers in Dornoch, my father was a dutiful son to his mother and combined seeing her with commitments in Hull, such as serving on the council of the university. Although he bore the brunt of the expenses of the Hessle household, Geoffrey contributed too. After lecturing in agricultural chemistry at Aberdeen and Leeds, he emigrated to Amani, Tanganyika (Tanzania) to investigate soils, a subject vital to a rural economy. At the colonial hill station, from where one could just discern the island of Zanzibar, he was surprised to find that one of his colleagues knew his brother. Harold Storey,[2] an authority on plant viruses, had belonged to the Cambridge University Natural Science Club, and my father, as secretary, had summarised his talk on cotton for the minute book.

Geoffrey unified soil cartography in East Africa. After classifying soils across Nyasaland (Malawi) and Northern Rhodesia (Zambia), he produced the definitive soil map of the region, and coined the term 'catena' for a sequence of soil types associated with topographical conditions. While applauding Geoffrey's work, my father cautioned him against getting bogged down in departmental affairs, and strongly urged him to publish his own research papers.

They kept in touch by letter. While my father liked gardening or taking a walk, his favourite form of relaxation was to write a letter. His tendency to write at inordinate length and in excessive detail prompted a well-meaning friend to suggest he refrain from such indulgence and use his time more effectively. But this took no account of its therapeutic effect on his state of mind, for in his letters my father released feelings that he would not have dreamt of expressing vocally, since social convention

---

[2] Harold Haydon Storey (1894–1960). Plant pathologist.

demanded that civilised people suppress their emotions. Indeed, revealing one's innermost feelings was interpreted as a sign of ill-breeding or want of manners.

After eighteen months in Africa, Geoffrey felt sufficiently secure in his job to marry and sent for his fiancée, Kathleen, a geography lecturer at Leeds University. My father had a great sense of family and saw her off from Tilbury on her bridal voyage. Every third year when they had home leave Geoffrey and Kathleen spent part of it with us. Meanwhile, decisions affecting Edith and Philip inevitably devolved upon my father.

Philip was a worry and a drain on the family's resources. He had the misfortune to graduate from Manchester University at the worst of the economic depression and failed to find employment. His luck changed during an extended visit to Germany, where his Hepworth talent for photography blossomed and he won several prizes. On returning home he got a job investigating insects in the upper air with the zoologist Alister Hardy,[3] an accomplished watercolourist, who had studied whales on the 1925 *Discovery* expedition to the Antarctic. Having turned entomologist, Philip specialised in the diseases of bees and he became advisor in beekeeping to the Ministry of Agriculture at Rothamsted Experimental Station, Hertfordshire. (He made mead and honey from the hive in his garden.) Responsible for the health of the nation's bees, he was occasionally summoned to the House of Commons to report on their welfare. So despite Edith and Sidney's lack of interest in science, each of their sons found his own niche.

This diversion races ahead of the narrative. By the time of the move to 19 Northmoor Road my parents were well settled into Oxford, the strands of professional and private life interwoven, but with one sorry difference from their Manchester days. There both my parents were equally part of the social scene, whereas at Oxford my mother was cut off from my father's life at Wadham. The neighbourhood was friendly and she knew many families, yet the contrast must have disappointed her. Oxford University was emphatically masculine and its ancient colleges were bastions of male supremacy, each a little world of its own, akin to a monastic

---

[3] Sir Alister Clavering Hardy (1896–1985) held professorships at Hull, Aberdeen and Oxford. He founded the Religious Experience Research Centre.

community, self-contained and self-regarding. Wives were perceived as superfluous appendages, and generally excluded. It mattered not whether a Fellow were married; he was still expected to dine regularly. The physicist Erwin Schrödinger thought Oxford was barbaric for the way it treated women. He disliked dining in college because of the lack of mixed company, and this, together with his complicated private life, influenced his decision to return to Austria. Only in a few rare instances, such as when Wadham entertained guest couples from overseas, was my mother included, and then in order to help entertain the wives.

When people came to stay, my father customarily took the man of the party to dine and savour the distinguished company of Wadham's High Table. The Dean and future Warden Maurice Bowra,[4] short and stocky, a master of pun and parody, was famed for his cauterising if sometimes malicious wit. David Cecil,[5] slim, gentle, boyish-looking with his delicate long thin face, had a supreme gift for the art of conversation. On one formal occasion Jean's husband Edwin was amused to detect David Cecil's white tennis shoes peeping out beneath his black dress trousers.

After my father's initial brush with Warden Stenning, they got along sufficiently well for my father to take him to a commemoration dinner at Trinity College, Cambridge. Checking the seating plan he was aghast to see that Stenning was not on the High Table, as befitted the head of an Oxbridge college. Having put this right, my father took action to save the Master from embarrassment, as, at such short notice, he might find himself at a loss for words to toast his Oxford guest. With profuse apologies for his impertinence, my father reminded the Master of the historical links between the colleges. Christopher Wren, the architect of Trinity's famous library, was a Wadham undergraduate; John Wilkins, Warden of Wadham, became a Master of Trinity; and Richard Bentley, Master of Trinity, who built the observatory on the top of Trinity's Great Gate, had studied at Wadham.[6]

---

[4] Sir Maurice Bowra (1898–1971). Classical scholar. Warden of Wadham from 1938 to 1970. Vice-Chancellor from 1951 to 1954.

[5] Lord Edward Christian David Gascoyne Cecil (1902–1986). Man of letters. History tutor at Wadham from 1924 to 1930. Professor at Oxford from 1949 to 1970.

[6] Letter from Milne to J.J. Thomson, 12 March no year (c. 1937), papers of J.J. Thomson MS.Add.7654, Cambridge University Library.

My father acquired a warm affection for Wadham. He was proud of its early role in nurturing scientific enquiry; he liked reeling off the heraldic description of the college's coat of arms; he was versed in the pleasing symmetry and homogeneity of its architecture,[7] hardly altered since Somerset masons completed it in 1613; and he was familiar with Wadham's trees, some of them specimens of international importance.

Wadham, neither rich nor fashionable, was regarded as a poor man's college, but it was a college on the rise. Small and friendly, with only nine Fellows,[8] its undergraduates were not split into factions of hearties and aesthetes. Bowra encouraged tolerance and kept rules to a minimum. On Bump Supper Night[9] he cunningly limited the boat club's revelry and prevented its wild antics getting out of hand and resulting in damage to college property by hiding under the bushes cheap second-hand furniture for them to 'discover' and hurl on their bonfire. He managed college business adroitly and briskly disposed of unwanted comment, yet, according to a contemporary,[10] if Milne stood up to Bowra with a few words, Bowra listened to him and respected his judgement. Under Bowra's flamboyant personality and panache Wadham grew in stature.

Milne's lack of bluster and unassuming manner suited the college's unpretentious ethos, and the Fellows, mostly historians, classicists and lawyers, accepted him. Certainly he could hold his own in conversation and few could surpass his knowledge of church ritual, architecture and history. He struck up a friendship with the scholarly young chaplain, Humphry House,[11] who admired Milne's capacity to reach out to others and take an interest in their subjects, although they could not reciprocate and penetrate his. House, a deacon, was all set to be a priest but lost his

---

[7] Since Milne's day the college has added many buildings behind the main quad.

[8] Today there are sixty-two Fellows.

[9] Bump Supper Night celebrates the end of the week-long inter-college rowing races. Having started in order of merit, a boat moves up a place by 'bumping' the one ahead.

[10] Conversation on 3 April 1990 with John Bernard Bamborough (1921–2009). First Principal of Linacre College, Oxford.

[11] Arthur Humphry House (1908–1955). Literary critic and teacher. In 1948 he became a university lecturer at Oxford and Wadham welcomed him back to a fellowship.

faith and resigned. He was in his early twenties, much the same age as Milne was when he entered his agnostic phase.

The stalwart backbone of the college was the physicist, Thomas Keeley,[12] gruff and laconic, who invented the electrometer which bears his name. In an unlikely partnership with David Cecil he had re-organised the lighting in the hammer-beam hall. Wartime aeronautics had thrown Keeley and Lindemann together, and Lindemann entrusted Keeley with the day-to-day management of the Clarendon Laboratory and paid part of his salary. Together they modernised Oxford physics. Milne was on affable terms with Keeley, but, like Lindemann, he was first and foremost an experimentalist and wanted no truck with theoretical physics.

For seven years Milne was the sole Fellow in mathematics. When the sale of a tract of land to Rhodes House put the college finances in better fettle that enabled it to fund an additional fellowship, he leapt at the opportunity to gain a mathematician. He had his eye on Jack Thompson,[13] who had written his doctorate on a topic Milne had suggested in response to an enquiry from the great physiologist Sir Charles Sherrington.[14] (Earlier Thompson had studied with the quantum physicist Max Born[15] in a forest in the Italian Dolomites after Born was forced to flee from the Nazis.) Milne successfully persuaded the fellowship to appoint Thompson. In him Milne gained a mathematical colleague and Wadham gained an astute bursar, who went on to be Senior Proctor of the university and, unusually for one who was not a head of house, a Pro-Vice-Chancellor.

The Fellows were a closely-knit little band, not always of one mind, but conscientious in putting first the wellbeing of the college. With scant clerical assistance, they exercised tight control over every aspect of the college. They handled its finances, maintained the buildings and gardens, supervised the kitchens and library and checked the college silver. After

---

[12]Thomas Clews Keeley (1894–1988) lived in Wadham for sixty years.

[13]John Harold Crossley Thompson (1909–1975). His doctorate was on the thermodynamics of elastic material like muscle. Bursar of Wadham from 1947 to 1964.

[14]Sir Charles Scott Sherrington (1857–1952). Professor at Liverpool from 1895 to 1913 and at Oxford from 1913 to 1935. President of the Royal Society from 1920 to 1925. Nobel Laureate 1932.

[15]Max Born (1883–1970). Tait Professor at Edinburgh University from 1936 to 1953. Nobel Laureate 1954.

discussing whether to buy a motorised lawn mower, they trooped out together to inspect the new-fangled machine. That a married Fellow might have other calls on his time mattered not a whit, and raises the question as to the extent to which the precedence of college duties might have stirred resentment at home.

Chapter 11

Cosmic Inspiration

Cosmologists are often wrong, but never in doubt.[1]
It is an essential requirement for a theory to stick out its neck.[2]

Six weeks after moving house, the Milnes' German friends, Kate and Erwin Freundlich,[3] whom they had met in Leiden, came to stay. The husbands were of one mind on the importance of a broad approach to astrophysics and astronomy. As Milne put it, 'The history of theoretical astronomy shows that specific theories of the constitution of celestial bodies have in general been less fruitful than investigations of fundamental problems associated with idealised models.'[4]

Freundlich, tall, forthright, lively, sporting a leonine crop of hair, was a notable exception to the post-war ban excluding Gemans from international scientific gatherings, and it was by personal invitation that he had participated in the Leiden IAU. He was an early champion of Einstein's theory of general relativity and among the first to understand its implications. To test its predictions, Freundlich undertook many eclipse expeditions to measure the deflection of starlight caused by the Sun's gravitation. Freundlich discussed his evidence at Milne's colloquium and Milne presented it to the RAS.

Freundlich was Director of the Einstein Institute[5] in Potsdam, which he founded with the express purpose of strengthening the empirical

---

[1] Attributed to Lev Landau.
[2] Bondi. H. *Assumption and Myth in Physical Theories*, Cambridge: Cambridge University Press, 1967, p. 1.
[3] Erwin Finlay Freundlich (1885–1964). Forced to leave Germany in 1933, he built up centres of astronomy at Istanbul, Prague and St Andrews, Scotland.
[4] 'The Radiative Equilibrium of a Planetary Nebula' in *Zeitschrift f. Astrophysik* 1 (1930), p. 98.
[5] Later renamed the Institute of Solar Physics and incorporated into the Potsdam Observatory.

foundations of Einstein's theory, and which he built with money he raised from industry. Yet hidden behind this remarkable achievement lay an unsavory tale of prejudice and persecution. For some years Freundlich had worked at the long established Potsdam Observatory under its rigidly conservative director Professor Hans Ludendorff,[6] brother of the World War I General Erich Ludendorff. Its staff concentrated on classical astronomy, recording the positions, brightnesses, motions and distances of individual stars, and spurned modern investigations into the chemical and physical nature of stars. Riddled with anti-Semitism, the observatory condemned Einstein's ideas, so that Freundlich's persistence in backing them, even to the point of deliberately publishing an account of general relativity against the wishes of the director, precipitated a crisis. Freundlich was ostracised and he resigned. That he was able to patch up personal relations with the observatory to build his institute within its grounds says much. Having gained the freedom to run his own institute, he used his excellent international contacts to attract to it people of the highest calibre.

While Milne was in Berlin the previous summer, he had gone to see Freundlich and his colleagues and they gave him the warmest of welcomes. Now Freundlich invited Milne to spend some time with them. Given what Freundlich had endured during his academic isolation, he probably sensed Milne's frustration at Oxford, albeit from an entirely different cause. Certainly they shared the quality of strong conviction even if it meant going against the grain of the majority. Milne was eager to accept the enticing prospect of working at Freundlich's institute but feared that Oxford University might refuse him leave of absence, something not lightly granted, unlike at Cambridge where Ralph Fowler and Paul Dirac frequently disappeared to Copenhagen to confer with Niels Bohr. Milne was relieved to be granted permission and the Rockefeller Foundation offered to fund him for the whole year, but as it would be Plaskett's first year in Oxford, Milne confined his visit to the autumn term.

---

[6]Hans Ludendorff (1873–1941). Director of the Potsdam Observatory from 1921 to 1938. Despite his Nazi tendencies, he protected Karl Schwarzchild's half-Jewish son in World War II.

The Potsdam plan wiped out Milne's previous plan to return to America for the IAU conference in Cambridge, Massachusetts, and then go west to Mount Wilson. With many regrets Margot declined an invitation from the Thompsons. Chandra had expected to join Milne in Oxford that autumn so Milne suggested he join him at Potsdam, but Chandra chose Copenhagen instead. Milne did arrange for two of his Oxford DPhil students to study in Germany.

At this point Milne was engrossed in the ramifications of stellar structure and felt that the subject warranted a book, though 'time always escapes me',[7] he told H.N. Russell. A reasonable conjecture is that Milne intended to write the book at the Einstein Institute, away from the everyday distractions at Oxford. During the spring of 1932 he had speaking engagements at the Durham Philosophical Society and the Leeds Astronomical Society, whose fortunes were revived by the energetic Selig Brodetsky,[8] and he made overnight trips to Cambridge and Manchester. In June he collected an honorary degree from Amsterdam University, which underlined his rapport with European astronomers, who tended to be better disposed towards his work than his British colleagues. The degree was probably due to Anton Pannekoek, founder of Amsterdam's Astronomical Institute, who, when in Lapland for the 1927 eclipse, confirmed certain predictions Milne had made about the photosphere. Another honorand was the small, frail Italian Tullio Levi-Civita,[9] whose mathematics paved the way for tensor calculus, essential to the study of relativity, and today relevant in the study of 'wormholes'.

Usually Milne carried through his intentions but on this occasion he never even began his book because by May he had swerved off in an entirely new direction. A casual remark by Jeans catapulted him into cosmology, a subject that may be defined as offering description and

---

[7] Letter from Milne to H.N. Russell, 3 July 1931, papers of Henry Norris Russell, Manuscripts Division, Department of Rare Books and Special Collections, Princeton University Library.

[8] Selig Brodetsky (1888–1954). Mathematician and Zionist. Professor at Leeds from 1924 to 1948. President of the Hebrew University at Jerusalem from 1949 to 1952.

[9] Tullio Levi-Civita (1873–1941). Professor at Padua and Rome. Modern applications of his work are described in *New Scientist* (23 March 1996), p. 30.

understanding of the structure and evolution of the Universe. It was to dominate his thinking for the rest of his life.

Jeans was at the pinnacle of his powers. Between 1929 and 1934 he averaged a bestselling book a year and entranced the public with his broadcasts. In a lecture at Manchester weighing up the merits of rival theories about the Universe he stated that we needed to know more about the disturbance which triggered its expansion.[10] His complacent prediction that such information was 'probably forever beyond our reach'[11] touched a raw nerve in Milne, who remonstrated that it was 'carrying scientific pessimism too far'.[12]

Milne felt driven to counter Jeans's gloomy defeatism and ten days later he had an answer. During this short period of gestation he gave his Halley Lecture on white dwarf stars without a hint that he was in the throes of 'vastly exciting'[13] mental activity. 'Like the flinging back of a curtain'[14] an image of the infant Universe had flashed before him that provided a reason for the expansion. If Jeans sparked Milne's imagination, Einstein was his inspiration and Milne was thrilled to explore a fresh field unshackled by the clutches of existing orthodoxies. For some years he had ruminated on Einstein's theories and hailed general relativity as 'one of the greatest triumphs of evidence for sheer intellect the world has ever seen'.[15] Uppermost in Milne's mind, however, after seeing him in Oxford and Berlin, was Einstein's special theory of relativity.

Solutions to Einstein's field equations had spawned a crop of cosmological models, although none of them addressed the cause of the expansion. Interpretations by Einstein, De Sitter, A. Friedmann, G. Lemaître and others, gave rise to Universes that were variously open or closed,

---

[10]Later coined 'the big bang' by Sir Fred Hoyle.

[11]Jeans's Ludwig Mond Lecture on 9 May 1932, reported in *The Times* on 10 May, prompted letters on 14, 18, 21, 23, 24, 25 and 27 May.

[12]Milne, E.A. *Relativity, Gravitation and World Structure*, Oxford: Clarendon Press, 1935, p. 3.

[13]Letter from Milne to H.N. Russell, 3 June 1932, papers of Henry Norris Russell, Manuscripts Division, Department of Rare Books and Special Collections, Princeton University Library.

[14]Letter from Milne to G. Milne, 10 August 1932, EAM, MS.Eng.misc.b.423.

[15]Letter from Milne to W.G. Thompson, 22 November 1930, in the author's possession.

static, expanding or contracting, some filled with matter and some empty. They depended on the choice of curvature of space, whether negative, flat or positive, and on the value ascribed to λ, Einstein's elusive 'cosmological constant', whose significance is undergoing a renaissance.

Milne's theory was in a completely new style which was quite unlike all previous theories. Whereas previous models concentrated on space and geometry, Milne wove his around light and time. In his mind he envisaged the infant Universe as a compact swarm of particles, buzzing like bees randomly in all directions at all kinds of speeds. Consonant with his Christianity, Milne credited the existence of the swarm to a creator. Always on the lookout for theological support, Milne gleefully noted that 'the earliest reference to the expanding Universe [is] in Genesis Chapter I verse 6, where according to the marginal note in the Authorized Version, the word "firmament" has the alternative rendering "expansion"'.[16] (Milne liked drawing on ecclesiastic architecture for similes. Kinematics is like 'a Gothic cathedral with its columns, tracery and buttresses. ... stained glass ... [gives] it variety, colour and individuality'.[17])

Milne replaced the idea of 'expanding space', current then as now, with his expanding swarm of nebulae whose boundary moves with the velocity of light. As the faster particles race ahead of the slower ones they form an ever-bigger spherical frontier, which Milne showed obeys Hubble's Law. (Any particle initially going towards the centre traverses it and moves outwards.) The swarm expands ineluctably and forever. With each particle Milne associated a galaxy and an observer, who considers himself to be at the centre of the Universe.

In both structure and substance Milne's theory was strikingly original. Regarding structure, scientific theories fall into two categories: inductive and deductive. Hitherto cosmological theories were inductive; they depended on extrapolating observations, i.e. going from the particular to the general. If an inductive theory has to accommodate anomalous results, then extra assumptions are tacked on piecemeal, a practice that Milne

---

[16]Letter from Milne to J.J. Thomson, 10 August 1932, papers of J.J. Thomson MS.Add.7654, Cambridge University Library.

[17]Milne, E.A. 'Remarks on the Philosophical Status of Physics' in *Philosophy* 16 (1941), p. 371.

scorned. His theory was of the deductive type, the best example being Euclidean geometry, and an ideal deductive theory, according to Milne, would be sufficiently robust and well crafted to allow for awkward results. He started from a few basic principles and systematically built up, step by step, a consistent theory, having pared his assumptions to a minimum — he put enormous store on the economy of Occam's razor to prune extraneous assumptions.

In substance, Milne was no less innovative. He dispensed with mass, energy and gravitational force, the hallmarks of dynamics, and focused on motion, later naming his theory 'kinematic relativity'. The essence of his thinking was that we observe the motion of bodies, not their forces. As he was fond of pointing out, it was the motion of the planets that prompted Kepler to formulate the laws which describe their orbits.

Milne's approach was totally at odds to general relativity, which firstly lays down the laws of gravitation that in non-relativistic cases reduce to Newton's, and secondly chooses an appropriate form of space–time to comply with known conditions in the Universe. He adopted the reverse procedure by specifying a type of space–time and then choosing a law of gravitation that produced the observed Universe.

Borrowing from Einstein's special theory of relativity, Milne's first postulate was that the speed of light has a fixed value, regardless of whether the light emanates from or is received by a moving body. His second postulate concerned the macroscopic view of the Universe: that it contains no preferential place, direction or frame of reference. While crediting Einstein with this, Milne extended it to include an observer and stated that the Universe must *appear* to be the same to all observers wherever they are. This 'cosmological principle', a cornerstone of many mainstream cosmological models, is often attributed to Milne. The introduction of an observer was a conspicuous advance. As Sir Joseph Larmor told Milne, 'You are the first person I have come across who pays any attention to the claims of the observer.'[18] It may seem counter-intuitive to regard the observer as an integral part of a scientific phenomenon, but this concept was gaining credence and permeates quantum theory.

---

[18]Letter from J. Larmor to E.A. Milne, 18 July 1932, in the author's possession.

Our awareness of the flow of time, the 'before', 'now' and 'after' of events, underpinned Milne's theory, and he designated time a fundamental unit, giving it precedence over space. 'To discuss time is to discuss the core of experience. To discuss space is to discuss the scaffolding of a building ... But the scaffolding is arbitrary. ... What is important is what is inside the scaffolding, the structure itself.'[19]

By assigning time a central role he broke with the usual practice of treating distance as a fundamental unit. He replaced the 'rigid ruler' for measuring distance, a notion highly unsatisfactory to Milne, with a 'clock', which he defined as a device for recording a sequence of temporal events at a given place. For him, distance was of subsidiary importance, deduced by timing light-signals in a technique which foreshadowed radar. If we note the time that elapses between an object emitting a light pulse to another object and receiving it back by reflection, a simple calculation gives the distance between the objects, provided the speed of light takes a fixed value. This, of course, is the method by which radar locates an object such as a submarine. Milne was mocked for the absurd impracticability of measuring cosmic distances in this way, though as soon as radar made its appearance he was censured for failing to spot its practical application. Eventually the clock and light-signal method for quantifying distance overtook the 'rigid ruler' approach. Today the distance measure of a 'light year' is common parlance.

In May 1932, after fulminating about Jeans's pessimism, Milne invited his students at his colloquium to get up any relevant literature, but nobody came forward,[20] and Milne said he would talk about the expanding Universe on Tuesday 7 June. The students expected a survey of recent work and were taken aback to hear him propound his own theory. He took care to apprise Einstein of the vital part played by special relativity and sent him an advance copy of the article about his theory that appeared in *Nature*.[21] Einstein thought Milne had a brilliant mathematical mind but lacked critical judgement, and while he did not

---

[19]Milne, E.A. *Relativity, Gravitation and World-Structure*, Oxford: Clarendon Press, 1935, p. 289.

[20]G.J. Whitrow's notebook, p. 3, in the author's possession.

[21]*Nature* 130 (2 July 1932), pp. 9–10.

approve of the theory to any great extent, he replied politely that it was plausible.

The concept of a fixed, static, unchanging world was deeply embedded in the public's subconscious and people were slowly adjusting to the evolutionary nature of the Universe. New contributions were welcome and in a burst of attention, news of Milne's theory spread from Budapest to California. It caused a flurry of excitement because it seemed to offer a strong alternative to Einstein's general relativity and it attracted the notice of physicists, historians of science and philosophers. The great twentieth century philosopher Karl Popper,[22] who followed its fortunes, commented that cosmology 'is a largely speculative subject … to have interesting ideas is much'.[23]

Science has its fashions, and during the inter-war years simplicity and elegance were in vogue. Milne's theory scored well on both counts. Like H.N. Russell, he disliked curved space and a subsidiary purpose of his proposal was to show that a model in flat space is feasible. Milne was elated by his straightforward, commonsense explanation and argued that its intrinsic simplicity gave it the edge over theories cluttered with elaborate notions of curved space and complicated geometry. He felt it was 'so beautiful that it must be true'.[24]

The bold departure of kinematic relativity from convention provoked hostile comment, but even its detractors admitted that inventing it was a feat of considerable ingenuity. Milne earned plaudits for introducing two useful terms which differentiate between how the Universe appears to an observer at some particular instant ('world picture') and how it really is ('world map'). These are not the same. His 'world picture' is what we see, a collage of images of celestial objects as they were in the past at various times — the further away in space, the earlier the time – as light has to travel to reach our eyes. (Sunshine takes eight minutes to reach the Earth.) Milne's 'world map' denotes the Universe at a particular moment in its history.

---

[22] Sir Karl Raimund Popper (1902–1994). In 1937 he left Austria for New Zealand and in 1946 came to England. Professor at London University from 1949 to 1969.
[23] Letter from Sir Karl Popper to the author, 9 December 1992.
[24] Quoted by G. Tyson in a letter to the author, 15 March 1991.

Many astronomers who admired Milne's pioneering work in stellar atmospheres regretted his veering off on a cosmological tangent and they regarded kinematic relativity as a quirky whim. Physicists, Max Born for example, dismissed it as a weird fantasy lacking connection with reality.[25] Besides, it offered little advance on previous theories when certain results from kinematic relativity were found to tally with a limiting case of Einstein's theory. Kinematic relativity lost further credence through its failure to produce an adequate theory of gravitation, although Milne never tired of wrestling to find one. He stubbornly ignored points raised in criticism, whatever their validity. Perhaps he was haunted by the emotional toll of his spat with Eddington, and wished to avoid more argument.

The chief impact of kinematic relativity was to enrich cosmology. The wealth of discussion it generated influenced future developments and formed the background to the steady-state theory proposed in 1948 by Hermann Bondi[26] and Thomas Gold,[27] later abandoned by them, and by Fred Hoyle. Attacks on the foundations of kinematic relativity led to the scrutiny of the bases of other theories and sparked curiosity about the enigmatic nature of time. Had kinematic relativity survived beyond its heyday it would have been dubbed 'revolutionary'[28] in a complimentary sense.

Throughout the summer Milne excitedly thrashed out the mathematics which gave him the number of galaxies in the Universe and its density. He stumbled on the hard four-dimensional geometry and made trivial slips. 'My algebra is infested with errors.'[29] He was all agog to get to Potsdam 'in the heart of Einstein territory'.[30]

---

[25] Born, M. *Experiment and Theory in Physics*, Cambridge: Cambridge University Press, 1943, p. 41.

[26] Sir Hermann Bondi (1919–2005). Mathematician. Chief Scientific Advisor to the Ministry of Defence from 1971 to 1977. Master of Churchill College, Cambridge, from 1983 to 1990.

[27] Thomas Gold (1920–2004). Scientific polymath. In 1956 he left England for the USA and held appointments at Harvard and Cornell.

[28] Harder, A.J. 'E.A. Milne, Scientific Revolutions and the Growth of Knowledge' in *Annals of Science* 31 (1974), pp. 351–363, especially p. 352.

[29] Letter from Milne to J. Larmor, 20 July 1932, MS 603 -613, at XXVIII.6, 1432, Royal Society, London.

[30] Letter from Milne to J. Larmor, 24 August 1932, MS 603 -613, at XXVIII.6, 1433, Royal Society, London.

On 2 September he left England, detouring via Zurich for a mighty gathering of some six hundred mathematicians from seventeen countries at the International Congress of Mathematicians. The proceedings in German were good preparation for the months ahead. In Potsdam the Milnes had a comfortable flat among chestnut and oak trees and when setting off for an evening with friends they took torches to wend their way along the dark paths through the woods. Just as America had been a release from Oxford's barren landscape, so too was Potsdam. My father treasured the months among the hospitable community of astronomers.

The Einsteinturm (Fig. 15), set in a clearing on Telegraph Hill above the grand palaces of Frederick the Great, is a bizarre and impressive sight. Now a listed building, the curving white tower, seven storeys high, is topped with a dome for a solar telescope. Its futuristic style[31] symbolised the new astronomy to make a deliberate contrast with the patterned brick of the old Potsdam Observatory. The inside is clad in wood and the double walls of the underground spectroscopy laboratory kept it at a constant temperature.

Before writing up his theory of kinematic relativity, Milne completed two papers on stellar energy[32] and he gave lectures which provoked 'exciting' discussions.[33] Nothing unusual in that. He was in high spirits and wrote:

> I have enjoyed being here most tremendously. … I have immediately set to work (a) to learn to speak German, (b) to work out my relativistic kinematic view of the Universe. I have found the former harder than the latter. …
>
> Two days ago I gave an account of my relativity theory here in German (or such German as I could muster), and Laue,[34] Schrödinger … and others

---

[31]The Einsteinturm was designed by the expressionist architect Erich Mendelsohn (1887–1953).

[32] 'Notes on the Boundary Temperature of a Star' in *Zeitschrift f. Astrophysik* 5 (1932), pp. 328–336; 'The Theory of Stellar Structure II (Energy-generation)' in *Zeitschrift f. Astrophysik* 5 (1932), pp. 337–347.

[33]Letter from H.A. Brück to the author, 26 January 1995. Among his Potsdam colleagues were Professor Walter Grotrian (1890–1954) and Dr Harald von Klüber (1901–1978).

[34]Max von Laue (1879–1960). Professor at Berlin from 1919 to 1943. Nobel Laureate 1914.

came over from Berlin, ten Bruggencate[35] from Griefswold ... Heckmann[36] and [others] from Göttingen. Ludendorff also turned up. I doubt whether any of them understood what I was driving at, except Schrödinger, who is very much on the spot. ... I have carried the argument enormously farther than when I last saw you ... and it comes to about 120 typewritten pages.[37]

Milne knew his paper[38] was too lengthy for the RAS and placed it with the leading European publication *Zeitschrift für Astrophysik*, which had already taken some of his work. (Later Milne was one of its editors.) Freundlich summarised the main points in German.[39]

While in Potsdam the Milnes visited the university town of Jena, a hundred miles away, where they saw the Zeiss optical works, the observatory and the planetarium. As Milne boasted to Geoffrey, he did not lack invitations:

I have just returned from visits to Munich and Göttingen, in each of which places I was handsomely entertained by the mathematicians and physicists concerned. I gave a lecture in each place in German. (I have already given 2 other lectures in German in Potsdam.) At Munich I saw the Deutsches Museum (equivalent to South Kensington), the Pinakothek, and the Munich Hof-Bräu or beer house. I swilled as much [beer] as was compatible with lecturing. I was given lunch also by Sommerfeld,[40] of atomic-theory fame.

---

[35]Paul ten Bruggencate (1901–1961). Director of Göttingen University Observatory from 1941 to 1961.

[36]Otto Heckmann (1901–1983). After working at Göttingen, he was Director of the Hamburg Observatory from 1941 to 1961.

[37]Letter from Milne to F.A. Lindemann, 3 November 1932, Cherwell papers, Nuffield College, Oxford.

[38]'World-structure and the Expansion of the Universe' in *Zeitschrift f. Astrophysik* 6 (1933), pp. 1–95.

[39]'Ein neuartiger Versuch von E.A. Milne, das kosmologische Problem zu lösen und die Expansion der Spiralnebel zu deuten' in *Die Naturwissenschaften* 21 (27 January1933), pp. 54–59.

[40]Arnold Sommerfeld (1868–1951). Physicist. Professor at the University of Munich from 1906 to 1947.

At Göttingen I saw Max Born, Herman Weyl[41] … and the hydrodynamical authority Prandtl[42]…

My work whilst here has been quite the most mature I have done … The problem of the gravitational motions in the Universe is exactly similar to that which faced Kepler in the late 16th century: what is the simplest description of these motions? … The solution … goes far to solve the classical problems of time and space, the infinity of space, whether time had a beginning, whether 'creation' has a meaning.[43]

Milne was disappointed by his discussions with Einstein. Nonetheless Milne 'extracted an admission from him that [kinematic relativity] is not the nonsense he once thought it was'.[44] He found Einstein 'not easy to talk with … [as he] interrupts before the point is reached'.[45] Exasperated, Milne resorted to writing.

You were kind enough to ask me some questions about my new work on world structure. … I simply imagine myself studying the distribution of matter and motion in the world, without any knowledge of physical laws except the constancy of the observed velocity of light … From the resulting kinematic picture I get dynamical and gravitational laws. You remarked to me that it was impossible to get dynamics from kinematics. Surely this is not true. Kepler studied the solar system, and tabulated the kinematical laws governing the solar system. From there, Newton, by a stupendous piece of insight, obtained dynamical and gravitational laws. I am trying to study the whole Universe instead of the solar system.[46]

---

[41]Hermann Klaus Hugo Weyl (1885–1955). Mathematician. Professor at the Swiss Federal Institute of Technology at Zurich from 1913 to 1930, at Göttingen from 1930 until 1933, when he left Germany for the Institute of Advanced Study, Princeton.

[42]Ludwig Prandtl (1875–1953). Aerodynamicist. Built the first wind tunnel. Director of the Institute of Technical Physics at Göttingen.

[43]Letter from Milne to G. Milne, 27 November 1932, EAM, MS.Eng.misc.b.423.

[44]Ibid.

[45]Letter from Milne to J. Larmor, 18 January 1933, MS 603 -613, at XXVIII.6, 1434, Royal Society, London.

[46]Letter from Milne to Einstein, 16 November 1932, by permission of the Hebrew University of Jerusalem, Israel.

Milne successfully canvassed Einstein to back Oxford University in its case against the Radcliffe trustees, and eventually persuaded Freundlich to do so too. After organising lawyers and affidavits to formalise their support Milne reported this progress to Lindemann in cheerful letters that radiate a confident sense of purpose.

In December Milne lectured in Copenhagen, the home of theoretical physics, stopping en route to give a seminar at the historic Baltic town of Greifswald. At Copenhagen he was the guest of the director of the observatory, Elis Strömgren,[47] and his wife. Milne was well acquainted with the work of their precocious son Bengt, who wrote his first papers in his teens, and who had recently confirmed that hydrogen was the dominating stellar element. Milne had helped him publish a paper with the RAS and was pleased to meet him and his wife. To help entertain Milne, the Strömgrens roped in Chandra, much the same age as Bengt, and one evening the conversation turned to the recent knighthoods for astrophysicists, Jeans in 1928 and Eddington in 1930, and likely future recipients. Under their jocular teasing Milne blushed crimson to his ears.[48]

My parents were in no hurry to leave Potsdam and for my two-year-old sister they dressed a Christmas tree in the German manner with wooden ornaments, which became precious souvenirs of a cherished sojourn that was never to be replicated.

For Freundlich and his institute it was also the end of an era. Once Hitler came to power, Einstein's name was obliterated from the institute and the staff were coerced into studying the effect of the Sun on radio transmission. Kate Freundlich was Jewish and she and Erwin fled the country. To Hermann Brück,[49] a likeable young man at the Einsteinturm, Milne's visit proved his salvation. When Brück could no longer stand the evil Nazi regime he took a temporary post at the Vatican Observatory on

---

[47] Svante Elis Strömgren (1870–1947). Director of the Copenhagen Observatory from 1907 to 1940.

[48] Conversation with S. Chandrasekhar on 24 October 1990.

[49] Hermann Alexander Brück (1905–2000). In 1947 he became Director of the Dunsink Observatory, Dublin. From 1957 to 1973 he was Director of the Royal Observatory, Edinburgh, and Astronomer Royal for Scotland.

a small grant. His desire was to reach England, and spotting an advertisement for the Radcliffe Travelling Fellowship, he wrote to Milne, the only person he knew there. Milne judged Brück a strong candidate and in a five-page letter encouraged him to apply, enumerating points he should make in his application.[50] Brück did not get the fellowship, but one thing led to another and he secured the post of Assistant Director at the Solar Physics Observatory in Cambridge. Like Milne, Brück enjoyed Newall's friendship, patronage and hospitality. Brück's distinguished career took off and ultimately he became Astronomer Royal for Scotland.

In March, during the same week that German elections increased Hitler's authority, Milne was in Aberystwyth to give public lectures on kinematic relativity on five consecutive afternoons. The week also saw President Franklin D. Roosevelt close American banks to stop a run on them, and a devastating Japanese earthquake cause 3,000 casualties. Disaster had struck Wales, too. Milne arrived in the aftermath of a terrible blizzard that derailed a train in a horrific landslide. Despite this, the captivating charm of Aberystwyth nestling in Cardiff Bay induced my father to bring us there – I was born in April – for an autumn holiday.

Milne's invitation to lecture at Aberystwyth was due to Thomas Lewis,[51] Head of the Applied Mathematics Department and a keen devotee of kinematic relativity. Milne was by far the youngest lecturer in a programme that included the archaeologist John Garstang,[52] the headmaster of Harrow Cyril Norwood[53] and Sir Arthur Salter,[54] who would become Member of Parliament for Oxford University.

In astronomical circles Milne was already well known, but his entry into cosmology propelled him onto a wider stage. He employed a press-cutting agency and Margot pasted into an album clippings that came from Britain,

---

[50]Letter from Milne to H.A. Brück, 15 December 1936; a copy is in the author's possession.

[51]Milne's obituary of Thomas Lewis (1898–1950) is in *Nature* 166 (19 August 1950), p. 296.

[52]John Garstang (1876–1956). Archaeologist. Professor at Liverpool from 1907 to 1941.

[53]Sir Cyril Norwood (1875–1956). Master of Marlborough College from 1918 to 1925. Headmaster of Harrow School from 1925 to 1934. President of St John's College, Oxford from 1934 to 1946.

[54](James) Arthur Salter, Baron Salter (1881–1975). Politician. Author. Member of Parliament for Oxford University, where he held a chair from 1937 to 1950.

America and far-flung colonial outposts. *John Bull* paid him the compliment of stating he was one of the younger generation who had made his mark since World War I.[55] Hull's press treated him as a Yorkshire lad made good. Manchester gave him friendly coverage. A photograph[56] shows him opening the physics laboratory at my mother's old school alongside Edward Fiddes, who drew applause by reminding the audience that my father had the good sense to marry a former head girl. A journalist commenting on Milne's explanation of the origin of cosmic rays in the *New York Times* summed up the current mood: 'Any Universe is only an hypothesis. Just now it is the fashion to design universes.'[57]

---

[55] '1918–1932: What has Youth Done?' in *John Bull* (6 August 1932).

[56] 'Withington Girls' School' in the *Manchester Guardian* (24 June 1933).

[57] A letter from Milne in *Nature* on 2 February 1935 prompted the article in the *New York Times* on 17 March 1935. The previous year, in February 1934, Milne had suggested that the origin of cosmic rays lay beyond the Solar System, a view that 'filled the bill' according to P.M.S. Blackett. His letter to Milne of 12 May 1934 is in the author's possession.

Chapter 12

Oxford's Enlightenment

The world as seen by science requires a creative mind.[1]

As the face of Oxford science began to change, the air sweetened and Milne felt less excluded, less downtrodden. The clearest evidence that the university had stopped ignoring science was that it designated a 'science area', carved somewhat grudgingly out of the Parks. An influx of talented Jewish physicists, recruited by Lindemann, revitalised the Clarendon Laboratory and in 1933 they were the first in the world to produce liquid helium. Early in World War II their groundbreaking research on the fission of uranium transformed the theoretical manufacture of an atomic bomb into a practical possibility.

The arrival of a trickle of Milne's friends who migrated to Oxford diminished his scientific isolation. Foremost among them was the engineer Richard Southwell,[2] whom Milne knew through Trinity and wartime aeronautics — Southwell served with all three armed services. He, too, had hesitated about taking up an Oxford chair, but, once ensconced, he scooped up former Cambridge colleagues and built a lively department. A man of notable charm, he and his tall, graceful wife Isabella raised a family of four daughters, whom they named after flowers. The arrival of the pre-eminent chemist Robert Robinson[3] and his wife Gertrude (also a chemist), who were friends from their Manchester days, added to the Milnes' circle. For Robinson, who made his name analysing and

---

[1] E.W. Barnes quoted in Barnes, J. *Ahead of his Age: Bishop Barnes of Birmingham*, London: Collins, 1979, p. 313.
[2] Sir Richard Vynne Southwell (1888–1970). Professor at Oxford from 1929 to 1942. Rector of Imperial College London from 1942 to 1948.
[3] Sir Robert Robinson (1886–1975). He held chairs at Sydney, Liverpool, St Andrews, Manchester, University College London, and Oxford. President of the Royal Society from 1945 to 1950. Nobel Laureate 1947.

synthesising naturally occurring materials such as dyes, Oxford was his sixth chair. Like the Milnes, the Robinsons believed in the importance of the social side of scientific life.

In mathematics, G.H. Hardy stole a march over Cambridge by grabbing for Oxford *The Quarterly Journal of Mathematics,* the successor to the *Messenger of Mathematics.* The principal editor was the pure mathematician Theodore Chaundy,[4] whose unrivalled knowledge of the printing of mathematics enhanced its stature. Considered an outstanding tutor, he had a profound sense of pattern, which drew him to botany and to folk dancing. He was one of the initiators of the May morning dancing on Magdalen Bridge.

Milne was always ready to contribute to *The Quarterly Journal* and its first issue opened with a paper from him on classical physics.[5] Generally he wrote on non-controversial aspects of physics and astrophysics. Take his two papers in December 1933. One[6] related to atomic collisions, inspired by Patrick Blackett's brilliant use of a cloud chamber to obtain visual evidence of positrons by photographing their tracks.[7]

The other paper,[8] twenty pages long, concerned Cepheid stars. Their regular variations in luminosity, arising from the rhythmic pulses of energy they emit, provide a standard for measuring the distance to stars, since there is a simple relationship between luminosity and the period of variation. The gist of the paper fascinated Adriaan Wesselink,[9] a young astronomer at Leiden, because it touched on his area of research and he wrote to Milne. Milne's reply gave Wesselink the courage to continue with

---

[4]Theodore William Chaundy (1889–1966). Born and brought up in Oxford, he served Christ Church for more than half a century.

[5]Milne, E.A. 'The Motion of a Fluid in a Field of Radiation' in *The Quart. Journ. of Math.* 1 (1930), pp. 1–20.

[6]Milne, E.A. 'On a Mean Free Path Formula' in *The Quart. Journ. of Math.* 4 (1933), pp. 315–318.

[7]Blacket, P.M.S. and Occhialini, G.P.S. 'Some Photographs of the Tracks of Penetrating Radiation' in *Proc. of the Roy. Soc. A* 139 (1933), pp. 699–726.

[8]Milne, E.A. 'The Energetics of Non-steady States, with Application to Cepheid Variation' in *The Quart. Journ. of Math.* 4 (1933), pp. 258–277.

[9]Adriaan Jan Wesselink (1909–1995). He left Leiden in 1946 to join its southern outpost in Johannesburg. In 1950 he moved to the Radcliffe Observatory and in 1964 to an appointment at Yale.

his line of enquiry despite the icy disapproval of his supervisor Jan Woltjer,[10] who dismissed Milne's paper as 'nonsense'.[11] Woltjer refused to sit down with Wesselink and listen to his point of view and stood holding the door knob, impatient for him to leave, a telling example of how research may be at the mercy of personal prejudice and behaviour. Fortunately for Wesselink, Ejnar Hertzsprung,[12] soon to take charge of the Leiden Observatory, did not share Woltjer's vilification of Milne's paper. Wesselink persevered and his name lives on in the Baade–Wesselink method for finding the radius of a pulsating star.

At Oxford the most gratifying event of all, the result of persistent nagging, was the decision by the university to grant mathematicians a home of their own. The grandly named Mathematical Institute[13] consisted of six rooms at the north end of the first floor of the Radcliffe Science Library extension, opened by the Princess Royal on 3 November 1934. There was a room for each professor, a library, a common room and £60 a year for upkeep, but not a penny for secretarial assistance.

Milne was jubilant. He put enormous store by personal contact and here was the clearest acknowledgement that mathematicians needed to interact with each other. The institute instilled a sense of cohesive identity and the benefits of having a departmental centre, however modest, could not be overstated. Mathematics grew in popularity and among non-medical sciences the number of undergraduates reading the subject was second only to chemistry.

The completion of the refurbishment of the OUO[14] was further cause for celebration. On 11 June 1935 the Vice-Chancellor inaugurated the new vertical solar telescope, the first of its kind in England. Eddington graced the occasion with a keynote address and his sister Winifred stayed overnight with us.

---

[10] Jan Woltjer (1891–1961). Dutch astronomer.

[11] Letter from A.J. Wesselink to the author, 16 July 1990.

[12] Ejnar Hertzsprung (1873–1967), Danish astronomer, gives his name to the Hertzsprung–Russell diagram that relates stellar temperature to magnitude. Director of the Leiden Observatory from 1934 to 1945.

[13] In 1952 the Mathematical Institute moved to 10 Parks Road and in 1966 to its current purpose-built premises in St Giles.

[14] The OUO was closed in 1988 and today there is a dome atop the physics building.

Although Plaskett concentrated on solar research, he cast his net wide and his colloquia at the OUO on Wednesdays complemented Milne's on Tuesdays. Their close bond made for a powerful partnership between observational and theoretical astronomy. (Additionally, twice a week, in sessions not listed in the *Oxford University Gazette* Milne discussed ideas that interested him at 10 am, an hour that he jested was 'too early for serious research'.[15]) Young and enthusiastic, they had markedly different personalities. Whereas Plaskett, handsome, assured and a man of the world, could be an exacting and tough taskmaster, sometimes feared by his research students, Milne was never intimidating. Small, slight, lacking social poise, he was less formal, more approachable. Plaskett's transatlantic background coupled with Milne's links to European astronomy engendered an international outlook and permanently dented the university's insularity. Together they created a world-class centre in astrophysics (Fig. 16).

The colloquia that Milne had initiated in 1929 gave an extra edge to postgraduate studies in astrophysics. In those days research students had no structured programme and were often left to their own solitary devices for weeks on end. They wholeheartedly welcomed the colloquia which were unique in Oxford, not least because Milne operated an easy-going 'open house' policy. (By contrast, membership of the Alembic Club for chemists, for example, was by election only.) He was glad whoever came, including a handful of dons: Hardy, Plaskett, Townsend and occasionally Lindemann. Some twenty people would cram into his room in Wadham, sprawling over the furniture or sitting on the floor. At first he held the sessions in the late afternoon but after dinner proved better because the discussion could continue into the evening until closing with a cup of tea. He ran them without repetition, year in and year out, moving them to the Mathematical Institute after it opened.

Tuesday was the high point of Milne's week and that of his students. He was quick to grasp new ideas, which he collected 'like burrs',[16] to toss about and identify their implications. He positively encouraged argument and spirited discussion. When he had to go to a funeral he exhorted the

---

[15] Letter from T.L. Leigh Page to the author, 31 March 1989.
[16] Letter from M.M. Crum to the author, 13 February 1991.

student who stood in for him to be 'as hard, abstract and provocative'[17] as possible.

The colloquia made students feel that they belonged to the scientific community for, if something newsworthy had occurred, Milne put aside whatever was planned and talked spontaneously, holding them spellbound with his intimate knowledge of scientists and their work. On the death of Rutherford, Milne revealed that if Rutherford entered the Cavendish whistling 'Onward, Christian Soldiers' he was in a good mood.[18]

The best evenings of all were those at which Milne brought along an outside speaker. The students realised that their good fortune in meeting distinguished astronomers hinged on Milne knowing them sufficiently well to persuade them to speak to a relatively small gathering. Among his foreign guests, apart from those already mentioned, were the Frenchman Léon Brillouin,[19] the Americans Edwin Hubble and Henry Norris Russell, the Indian Megnad Saha and the German physicist Arnold Sommerfeld. From nearer home there were Chandra, Eddington, Hardy and William McCrea.[20]

These 'marvellous occasions' might contain an element of surprise since the students were not sure 'which visiting eminence would appear', recalled Reginald Jones,[21] whose outstanding work in intelligence and radar was to hasten victory in World War II. One evening, as the students waited in Milne's room, they heard 'footsteps coming up the stairs and a very American voice saying, "After you, Sir, always after you"'.[22] This was the voice of Henry Norris Russell, renowned for his incessant flow of

---

[17]Letter from Milne to A.G. Walker, 22 May 1935, Walker papers, Balliol College, Oxford.

[18]Kendall, D.G. 'Statistics, Geometry and the Cosmos' in Bondi, H. and Weston-Smith, M. (eds) *The Universe Unfolding*, Oxford: Clarendon Press, 1998, p. 111.

[19]In the 1950s Léon Nicholas Brillouin was Director of Electronics at IBM.

[20]Sir William Hunter McCrea (1904–1999). Irish-born astrophysicist. In 1936 he was appointed professor at Queen's University, Belfast, in 1944 at Royal Holloway College and in 1966 at the University of Sussex, where he established its Astronomy Centre.

[21]Reginald Victor Jones (1911–1997). Wartime Director of Intelligence at the Air Ministry, and the first scientist employed by MI6. Professor at Aberdeen University from 1946 to 1981.

[22]Letters from R.V. Jones to the author, 25 February and 4 March 1991.

conversation, whom they were expecting. To their astonishment, Albert Einstein walked in with him and impressed them with his detailed knowledge of topics they considered outside his own particular field.

An informal gathering was more to Einstein's taste than an impersonal lecture. He liked to smoke, and Milne placed tobacco on the mantelpiece over the coal fire, instructing the students to keep him well supplied, which they did, jumping up in turn.

Once when Einstein was talking — Lindemann usually acted as interpreter — peculiar scuffling noises from the dingy little coal store next to the fireplace mystified the students. Suddenly, with a dramatic rattle of the handle, the door burst open to reveal the extraordinary spectacle of a startled undergraduate, his face smudged with coal dust. Stumbling and blinking in the light, he thought he beheld an apparition until it dawned on him that the man before him really was Einstein in flesh and blood. He 'clapped his hand to his forehead, uttered, "My God"',[23] and bolted from the room. Not wishing to appear inattentive or disrespectful to their illustrious speaker, the students stirred not a hair and pretended nothing had occurred. Later all was explained. The porter customarily locked Wadham's main gate after dinner, and, if asked, would willingly open it on payment of a token fine of a few pence.[24] Out of sheer bravado undergraduates would sometimes risk climbing into the college and on this occasion the hapless man was exploring a new route. Having scaled the front wall, he shuffled along to a specified window to gain entry, but he miscounted and landed in the coal store.

Oxford's DPhil degree was a comparatively recent invention and few bothered to study for it, partly because of the expense. Today it is hard to appreciate the acute severity of the financial constraints placed on needy students who had the appetite and ability for further study. Government loans or grants did not exist, so that, no matter how deserving a student might be, unless he had access to private means, a doctorate was well nigh impossible.

---

[23]Letter from R.V Jones to the author, 25 February 1991.
[24]The fines were one penny till 10 pm, three pence from 10 to 11 pm, sixpence from 11 to 11:30 pm and one shilling from 11:30 pm to midnight. Letter from R.V. Jones to the author, 23 June 1997.

Obviously this made for extremely stiff competition for the sparse funds available. Across all disciplines there were at most three Senior Studentships in the university, and in mathematics one Senior Scholarship. Apart from the Harmsworth at Merton College and the Skynner at Balliol, no college supported postgraduate studies in astronomy and astrophysics. Any man, and invariably they were men, who carried off one of these precious awards was patently of outstanding calibre. In the period from 1932 to 1937 two of Milne's pupils won Senior Studentships and five won Senior Mathematical Scholarships.[25] That Milne was a magnet for the brightest of the bright did not pass unnoticed in other quarters of the university. He was now a person of standing, no longer relegated to its margin.

Having struggled himself to surmount financial hurdles, he had a natural empathy for the desperate predicament of aspiring students and did everything he could to help them. A case in point is Leslie Camm.[26] Having subsisted on undergraduate scholarships and gained a First in mathematics, he yearned to continue his studies, but, being penniless, he had abandoned all thoughts of a DPhil. He was disconsolately at home in Sheffield when he opened a letter from Milne that raised his hopes. Inside the envelope he found a notice about the £200 Skynner Studentship that Milne had cut out from the *Oxford University Gazette*. Milne pressed Camm to apply without delay. He secured the vital award and his doctorate led to his life's work in stellar dynamics.

Nearly half of Milne's postgraduate students were Rhodes Scholars from America, Canada and South Africa. That they boosted the numbers studying astronomy contributed materially to establishing Oxford as a centre for cosmic studies. There was another factor, too. Although Rhodes Scholars were graduates of universities in their own countries, they usually took undergraduate courses for an Oxford BA degree. The standard of Milne's

---

[25] Senior Mathematical Scholars: J.H.C. Thompson (1932), A.G. Walker (1934), G.J. Whitrow (1935), S.W. Coppock (1936) and G.L. Camm (1937). Additionally, Thompson and Whitrow held Senior Studentships, Walker and Whitrow Harmsworth Scholarships. I am indebted to Roger Hutchins for pointing out the preponderance of Milne's pupils among the recipients of these awards. For further detail see Hutchins, R. *British University Observatories 1772–1939*, London: Ashgate, 2008.

[26] George Leslie Camm (1914–2000). Reader in Mathematics at Manchester University.

Rhodes Scholars, however, was so far above the run of the mill that the university agreed to make an exception and to allow them to embark on postgraduate degrees, mostly doctorates. Moreover, several of them completed their doctorates in two years, which added to their lustre.

In one-to-one sessions with students Milne explained a trifle paternalistically how they should express themselves in clear, correct grammatical English; he forbade the word 'very', always demanding a meaningful substitute. This stood them in good stead, as they came to appreciate, whatever their walk of life. He expected the highest standards and 'set his students the good example of not easily giving up when there was a tough problem to be solved', remembered Gabriel Cillié.[27] Faced with the laborious job of processing observational data, he and Geoffrey Wiles,[28] a relation of Andrew Wiles who famously solved Fermat's Last Theorem, were grateful to Milne for procuring, at the university's expense, a Brunsviga calculating machine, which saved them hours of tedium. Milne urged students to publish their research without delay, not just to gain recognition but also for the benefit of others.

Milne extended a hand of friendship to students, especially those from overseas, and invited them home. Hardly a week passed without somebody coming to a meal. Rhodes Scholars tended to have distorted preconceived notions about Milne and were surprised on first acquaintance to find him quiet and shy, somewhat retiring, without a vestige of aggressive argument. His botanical knowledge was an eye-opener to Wiles, who came upon Milne planting bulbs. When he talked about hybridisation, Wiles realised that his 'thought-life' was not confined to stars.[29] An American, Ivan Getting,[30] who rated Milne a 'giant',[31] found him on his knees unblocking a drain and decided that he must be mortal after all!

---

[27] Gabriel Gideon Cillié (1910–2000). Astronomer. Organist and composer of church music. Professor of Mathematics at Stellenbosch University from 1939 to 1975.
[28] Geoffrey Gilbert Wiles (b. 1907). Lecturer in Physics at Witwatersrand University and the University of Cape Town.
[29] Letter from G.G. Wiles to the author, 28 April 1996.
[30] Ivan Alexander Getting (1912–2003). President of the Aerospace Corporation from 1960 to 1978. His impressions of Oxford are recorded in 'Opportunities for Research in the Physical Sciences at Oxford' in *The American Oxonian* 24 (1937), pp. 21–24.
[31] Getting, I.A. *All in a Lifetime*, London: Vantage Press, 1989, p. 52.

Milne had a gift for breaking the ice with young people. His long friendship with David Kendall[32] sprang from a chance encounter at the OUO when Kendall was in his first term. He was destined to be the British leader in probability theory, which he brought to bear on subjects as disparate as parish records, epidemics and Egyptian archaeology. At the age of eighteen he was equally keen on astronomy and pure mathematics, a rare combination, and Milne issued a standing invitation to come to tea 'any time I liked'.[33] Kendall attended Milne's lectures and soon published a paper in astrophysics. He was good in a domestic setting, as Milne noted in a testimonial, and chatted to us children — he would go on to father six — and helped us rig up patriotic bunting for the coronation of King George VI.

The majority of Milne's postgraduate students clambered up the ladder of academe. Others excelled in related fields. Ivan Getting became eminent in defence and his work on military satellites led him to develop GPS (the global positioning system). Stephen Coppock[34] rose to be a senior figure in weapon research at the Ministry of Defence. His high velocity anti-tank shells bombarded the enemy on D-Day. Laurence Lefèvre[35] became Head of Mathematics at Eton; he thought astrophysics was a suitable school subject.

Relevant academic posts were few and far between. Among Britain's twenty universities there might be one or, at most, two a year. Milne kept his ear out for openings and advised his students how best to present themselves. 'Dress yourself up as a geometer',[36] he advised one student. He wrote countless references, and, if he found himself in the awkward situation of supplying testimonials for competing candidates or of sitting on the selection panel, he was scrupulous in making this clear. Time and

---

[32]David George Kendall (1918–2007). Statistician. Lecturer at Oxford from 1946 to 1962. Professor of Mathematical Statistics at Cambridge from 1962 to 1979.

[33]Bingham, N.H. 'A Conversation with David Kendall' in *Statistical Science* 11, 3 (1996), p. 162.

[34]Stephen William Coppock (1911–1991). Principal superintendent of research at Fort Halstead from 1962 to 1972.

[35]Laurence Eustace Lefèvre (d. 1981). He left Eton for a lectureship at the University of British Columbia, Canada.

[36]Letter from Milne to A.G. Walker, 31 January 1936, AGW.

again pupils came back to him and as he had no secretarial help except Margot, he was obliged to keep careful track of their job applications. He put aside the first hour after breakfast for dealing with administrative matters.

That left the rest of the day for his own interests. He shrugged off criticism about kinematic relativity and noted that comparing it against the 'gold standard' of general relativity was a popular pastime:

> It is amusing to record that Einstein has tried to make out that my solution of the cosmological problem is identical with his and De Sitter's involving *flat* expanding space; that Eddington has said that it is identical with Lemaître's involving *spherical* expanding space; that Heckmann has said at one time that it is only De Sitter's original world (empty) and at another that it is expanding *hyperbolic* space. It cannot be all four.[37]

At the 1933 BAAS meetings there was a fervid debate on rival cosmological theories. Milne's exposition prompted Eddington to accuse him of 'wrecking relativity theory and knocking the bottom out of space'.[38]

Although Milne was wrapped up in kinematic relativity he kept other ideas on the boil. Having been snuffed out by Jeans regarding the fate of the Universe at the earlier BAAS meeting, Milne used a talk at Oxford to meet the doom-mongers head-on. His title 'World Without End', with its whiff of spirituality, caught the attention of the press in Canada, India and Hong Kong. Extravagantly he claimed that only when the Universe is observed does it have age and size. Each observer thinks he is at the centre of the Universe and sees creation continually taking place at its extremities. 'The World, though old at its centre, is ever young at the edge. ... The Jeremiahs are thus probably wrong, and the world's message is one of eternal hope'.[39]

---

[37] Letter from Milne to S. Rosseland, 2 April 1933, SR, RA/PA-1514.

[38] For example, *The London Evening News, The Nottingham Evening Post, The Edinburgh Evening News* on 12 September 1933, *The Times* and *The Birmingham Post* on 13 September 1933.

[39] 'World Without End' in *The Oxford Junior Scientific Club Transactions* (1933), p. 181.

Milne hammered away on the same theme in his Joule Lecture to the Literary and Philosophical Society in Manchester.[40] The remit of the prestigious lecture, which commemorates the great physicist James Joule, who identified the equivalence of heat and energy enshrined in the principle of the conservation of energy, suited Milne to perfection. The lecturer is expected to probe some aspect of energy. In his treatment of the Universe as a thermodynamic system Milne drew attention to the curious dichotomy of the role of time. On the one hand, according to the laws of thermodynamics, the irreversibility of heat transfer implies a passing of time, illustrated by stars using their energy to shine. On the other hand, the one-way characteristic of time, the 'arrow of time' is not embedded in equations which define theories of relativity.

The concept of time was uppermost in Milne's mind. He thought it was of the utmost importance and at the British Institute of Philosophy[41] pleaded for a more profound understanding of its nature. Keenly aware of rumblings against his unorthodox outlook and his obsession with deductive reasoning, he differentiated between 'world physics', based on experiment, and the 'abstract reality' of his cosmological model. He confided to Chandra that he was getting to be 'less of a theoretical physicist and more of a half baked philosopher'.[42] Milne preferred the old-fashioned term 'natural philosopher' rather than scientist. For him, like his heroes Newton and Descartes, science, theology and philosophy were all of a piece, untroubled by lines of demarcation. Milne was in thrall to the writings of Ralph Waldo Emerson, who interpreted science as a path to godliness, and applauded Emerson's definition of philosophy: 'Philosophy is the account which the human mind gives to itself of the constitution of

---

[40] Milne delivered the Joule Lecture on 27 February 1934. See 'The Expanding Universe as a Thermodynamic System' in *Mem. Manchester Lit and Phil. Soc.* 78 (1934), pp. 9–40. Among his predecessors were A.V. Hill and Eddington; among his successors E.V. Appleton and C.G. Darwin.

[41] Delivered on 17 October 1933. Milne, E.A. 'Some Points in the Philosophy of Physics: Time, Space and Creation' in *Smithsonian Reports* (1933), pp. 219–238 and *Philosophy* 9 (1934), pp. 19–38.

[42] Letter from Milne to S. Chandrasekhar, 27 September 1934, EAM, MS.Eng.misc.b.427.

the world.'[43] Both Newton and Emerson were Unitarians,[44] one covert, the other open, which may have appealed to Milne's general sense of heterodoxy, although he was never tempted away from the Church of England.

Milne was pleased that Chandra decided to join him in Oxford for the Michaelmas (autumn) term of 1933. Wishing to offer him some financial assistance, Milne went to the length of consulting the Warden of Wadham, but to no effect. As Milne ruefully explained to Chandra, Oxford professors had no access to research funds. Fortunately Chandra came from a well-heeled Brahmin family and had no financial constraints. Then, on the morning of his departure from Cambridge to Oxford, he learnt that Trinity had elected him to a fellowship, but he did not alter his plans. Intensely thrilled at Chandra's success, Milne cast his mind back to his own pivotal election that opened up his life in mathematics and rejoiced with Chandra in a nostalgic wallow of elation.

Chandra took wry delight in finding lodgings at 19 Southmoor Road for its symmetry with 19 Northmoor Road. He often came to our house and the friendship he forged with us outlasted my father's lifetime. Until he left for America in 1936, he made a habit of coming to Oxford once a term, and on occasion stayed the night with us. He showered us with presents: books, chocolates, brightly 'jewelled' maharani dolls and an exquisite ivory cribbage board.

Chandra singled out one episode as being typical of Milne's trademark originality in considering a familiar circumstance from a fresh angle. Sitting in Milne's study, Chandra saw propped on the mantelpiece the postcard he had sent depicting a boy standing on his head with the caption, 'I see things right side up'. Milne had responded, 'Simpletons who stand on their heads see the world from a different angle — but very often it is they who turn the world upside down.'[45] While musing on the preconceptions which cloud the minds of adults, in contrast to the minds of children which are uncluttered by prejudice, and who can thus see things 'right

---

[43]From Emerson's essay on Plato, 'Lecture II. Plato: or the Philosopher' in *Representative Men*, 1850. Quoted by Milne in a letter to the author, 30 October 1942.

[44]I am obliged to Léon Mestel for pointing out the Unitarian link.

[45]Card from Milne to S. Chandrasekhar, 20 June 1935, EAM, MS.Eng.misc.b.427.

side up', Chandra challenged Milne to expand upon his remark. Milne turned to statistics. The well-known 'method of least squares', which finds the best curve for an array of scattered points, estimates a set of quantities whose number is one *less* than the number of constraining equations. Milne commented that the converse is rarely considered. What, he asked, is the estimate of quantities whose number is one *more* than the number of equations? As an example, he cited measuring errors in the markings on a ruler. Later, Milne took Chandra to dine in Wadham and they parted. Next morning they met at 9 am, and as they walked to the railway station for Chandra's train, Chandra was astonished to learn that overnight Milne had covered some fifty pages trying to solve his own problem, though with scant success. For nearly forty years the problem lay dormant until conquered by mathematicians at Imperial College London.[46]

Asking and tackling new questions was meat and drink to Milne. During February 1934 he and William McCrea invented Newtonian cosmology. In a whirlwind of excitement they bounced ideas back and forth at a great rate, and, remarkably, within two weeks they completed work which remains of value. In their exemplary collaboration, underpinned by close friendship, some credit for the extraordinary speed of their research must go to the excellent postal services between Oxford and London which consisted of three deliveries a day. This enabled them to exchange ideas rapidly, confident that a letter posted in the morning would reach the other in time for his reply to arrive back by the late afternoon delivery.

At Cambridge McCrea had sat in the front row at Milne's lectures and soon after McCrea achieved a First in his Tripos examinations they happened to be in Trinity's library. McCrea was enormously gratified that Milne immediately came up to offer his congratulations. This sparked a flourishing friendship. They kept in touch by letter and at meetings of the London Mathematical Society, the Royal Astronomical Society and the (London) Physics Club. After McCrea took a post at Imperial College London, they saw each other more frequently and he and his wife Marian often entertained Milne to a meal.

---

[46]Cox, D.R. and Herzberg, A.M. 'On a Statistical Problem of E.A. Milne' in *Proc. R. Soc. Lond.* A 331 (1972), pp. 273–283 and Herzberg, A.M. 'On a Statistical Problem of E.A. Milne' in *Proc. R. Soc. Lond.* A 336 (1974), pp. 223–227.

McCrea was a key player in the fast growing field of cosmology and had pointed out that certain results from kinematic relativity were compatible with those in general relativity, a conclusion also reached by the American Howard Robertson.[47] Milne had explored whether classical Newtonian physics could produce equations compatible with the expansion of the Universe[48] and asked McCrea, 'Has anybody shown that it is easy to construct a Newtonian expanding universe?'[49] He gave details of a particular case, parabolic, showing the equivalence between certain Newtonian equations and their relativistic analogues. At the start of an exceptionally busy day McCrea stuffed Milne's letter into his pocket, and when he did get around to reading it he was sceptical. He replied to that effect, but during the course of the evening, he made some calculations that disposed him to change his mind and he hastened out with another letter to catch the midnight post. Letters and postcards hurtled back and forth,[50] and they extended the theory to elliptical and hyperbolic cases, which generalised their results. They showed that the curvature of space, whether positive, flat or negative, was irrelevant to the expansion of the Universe, which could just as readily be explained by classical gravitation dynamics as by relativity. Cock-a-hoop, Milne boasted, 'This is highly sensational. We have brought off one of the biggest scoops in the history of modern physics, unless I have overlooked something.'[51] Milne invited the McCreas to come to Oxford on Saturday 17 February for lunch, tea and supper, enclosing a list of trains from Paddington, and they wrote their paper[52] that day.

Since their mathematics avoided complicated tensor calculus, essential for general relativity, they were astounded that nobody else had previously

---

[47] Howard Percy Robertson (1903–1961). American physicist at the California Institute of Technology. During World War II he worked in intelligence and liaised with R.V. Jones.

[48] Milne, E.A. 'A Newtonian Expanding Universe' in *The Quart. Journ. of Math.* 5 (1934), pp. 64–72.

[49] Letter from Milne to W.H. McCrea on 6 February 1934, WHMcC.

[50] Milne sent letters or cards to McCrea on 6, 8, 10, 11, 11, 14 and 18 February 1934, WHMcC.

[51] Letter from Milne to W.H. McCrea on 11 February 1934, WHMcC.

[52] McCrea, W.H. and Milne, E.A. 'Newtonian Universes and the Curvature of Space' in *The Quart. Journ. of Math.* 5 (1934), pp. 73–78.

thought of doing it. Had Isaac Newton or anyone else carried it out, the notion of an expanding Universe need not have waited until the twentieth century. The likely reason is that during the intervening three centuries every thinking person took for granted that the Universe was fixed in space, unchanging and eternal.

Newtonian cosmology was more to the taste of mathematicians than physicists, who disparaged it as a cerebral exercise. Milne, though, was swept away with enthusiasm, and saw an 'endless vista of investigations ahead',[53] while McCrea was more circumspect. In the words of the Nobel physicist Willie Fowler,[54] Milne 'put up a good fight for Newton'.[55]

Milne was wary about asking the RAS to publish their paper because the society had just accepted two papers from him and he knew that it disliked an excess of papers from any one author. (Milne deplored a 'most scandalous'[56] attempt by the RAS to limit an author to 40 pages per annum.) He knew that the next edition of *The Quarterly Journal of Mathematics* was 'short of stuff'[57] and approached Chaundy, who passed the paper to a referee. He pronounced it 'delicious'[58] and the paper appeared in March.

Milne intended to be 'popular, plain and Newtonian'[59] when he presented their findings to a roomful of international astronomers. Prominent among them was Edwin Hubble, who was in England to give Oxford's Halley Lecture. An Anglophile since his days as a Rhodes Scholar, he was given to wearing brogues and speaking in clipped tones. (Two years later he came again to deliver the Rhodes Lectures.) He and his wife Grace made Oxford their base for their month-long visit, which got off to a felicitous start with the surprise announcement of an honorary doctorate — the university was eager to acknowledge his fame and claim him as one of its

---

[53]Letter from Milne to W.H. McCrea, 17 April 1934, WHMcC.

[54]William Alfred Fowler (1911–1995). American astrophysicist. Director of the Kellogg Radiation Laboratory at the California Institute of Technology. Nobel Laureate 1983.

[55]Fowler, W.A. 'The Age of the Observable Universe' in Bondi, H. and Weston-Smith, M. (eds) *The Universe Unfolding*, Oxford: Clarendon Press, 1998, p. 172.

[56]Letter from Milne to S. Chandrasekhar, 18 January 1933, EAM, MS. Eng. Misc.b. 427.

[57]Letter from Milne to W.H. McCrea, 1 March 1934, WHMcC.

[58]Postcard from Milne to W.H. McCrea, 11 March 1934, WHMcC.

[59]Letter from Milne to W.H. McCrea, 8 May 1934, WHMcC.

own. Together Knox-Shaw, Plaskett and Milne co-ordinated a round of social events, recorded by Grace as 'jolly' lunches, 'pleasant' dinners and tea with us.[60]

Oxford was now firmly on the international circuit and in May Milne was perpetually on the go. He welcomed Harlow Shapley and the Dutch astronomer Jan Oort,[61] tall and ascetic, who stayed at our house. Oort confirmed Shapley's measurements of the Galaxy that show that our Solar System is not at its centre. Later Oort proposed that particles beyond Pluto are a source of comets: the eponymous Oort cloud. With understatement Milne commented that life was 'pretty full':

> We had lunched out to meet [the Hubbles] the day before [his lecture] and dined in state at Rhodes House in the evening. Next day Shapley ... came to Oxford and I gave a dinner party at Wadham in his honour. On Thursday I showed Shapley the work on his subject I have in progress ... and presided in the evening at a meeting of the Mathematical and Physical Society here. ... On Friday [at the RAS] I opened a discussion on the expanding universe, with the startling announcement that 'curved' and 'expanding' space is all bunkum and that a thoroughly good solution of the cosmological problem could have been obtained by Newton. ... We had dinner at the RAS Club afterwards, with Hubble, Shapley (who were kept from quarrelling), Oort of Leiden ... [and] Lemaître as foreign visitors. I met J.C. Squire[62] ... and was driven back to Oxford by road 12–2 am. Next day I had to entertain the sub-faculty of mathematics at Wadham to lunch, and attend a business meeting of the faculty all Saturday afternoon. By way of recreation we played bridge in the evening. Today we went to a christening ... On Thursday I have to give a full dress lecture[63] to the London Mathematical Society. ... Meanwhile lectures here come round and have to be delivered.[64]

---

[60]Grace Hubble's diary, Hubble Tape 74, SHQP.

[61]Jan Hendrick Oort (1900–1992). Director of the Leiden Observatory 1945 to 1970.

[62]Sir John Collings Squire (1884–1958). Man of letters, whose parodies Milne admired.

[63]In 'World-Gravitation by Kinematic Methods' Milne suggested a source of cosmic rays. *Nature* (26 May 1934), p. 789.

[64]Letter from Milne to G. Milne, 13 May 1934, EAM, MS.Eng.misc.b.423.

As if to endorse the university's enlightenment and Milne's vastly more agreeable life at Oxford, he received astronomy's highest accolade. The Gold Medal of the RAS is open to all nationalities and it had not come to an Oxford man since Charles Pritchard[65] in 1886. Milne was awarded the medal in January 1935 for his work in stellar atmospheres. On the brink of his thirty-ninth birthday, he is one of its youngest recipients and I know it was the medal he prized above all others.

---

[65]The Reverend Charles Pritchard (1808–1893). Amateur astronomer. Headmaster of Clapham Grammar School from 1834 to 1862. Appointed Savilian Professor in 1870.

Chapter 13

The Pendulum and the Atom

The universe begins to look more like a great thought than a great machine.[1]
I strongly approve of the attempt to prove truths from first principles.[2]

Life seemed tame after the frenzy of Hubble's visit and Milne was out of sorts. His usual exuberance deserted him for a conference at St Andrews[3] in late July to which he took Margot and Eleanor. He despaired that his ideas were not being judged on their merits[4] and sensed that the audience was out of sympathy with them. Additionally he felt inhibited[5] by the presence of the elderly esteemed Willem De Sitter, sporting his neat Vandyke beard, who with Einstein had produced a cosmological model that was widely accepted as an accurate representation of the Universe. Milne disagreed with De Sitter over the notion of a horizon beyond which light cannot reach us, but did not wish to clash with him publicly. Privately he likened De Sitter's mind to 'the hide of a rhinoceros'.[6]

By the autumn Milne had finished the manuscript of his book *Relativity, Gravitation and World Structure*[7] and until it was published six months later he fretted about how it would be received. To compound his despondency he was saddened by the deaths of three men he venerated: his predecessors in the Beyer Chair, Sir Arthur Schuster and Sir Horace Lamb, and his Trinity tutor R.V. Laurence.

---

[1] Jeans, J.H. *The Mysterious Universe*, Cambridge: Cambridge University Press, 1930, p. 148.

[2] Leibniz, G.W. translated by M. Morris. *Philosophical Writings*, London: Dent, 1934, p. 113.

[3] Milne gave a course of lectures at the conference, held from 18 to 28 July, organised by the Edinburgh Mathematical Society. *The Mathematical Gazette* 18 (1934), pp. 248–249.

[4] Letter from Milne to S. Rosseland, 2 April 1933, SR, RA/PA-1514.

[5] Letter from Milne to S. Chandrasekhar, 27 September 1934, EAM, MS.Eng.misc.b.427.

[6] Letter from Milne to A.G. Walker, 5 September 1934, AGW.

[7] Milne, E.A. *Relativity, Gravitation and World Structure*, Oxford: Clarendon Press, 1935. The publication date was 21 March 1935.

Milne feared that at worst his book would be ignored as the work of a crank, and at best it was bound to arouse hostile criticism.[8] He had no qualms about burdening his friends with his misgivings and warned Henry Norris Russell that the book was crude, outspoken and not scholarly.[9] To Chandra he wrote,

> I shall be told that I do not understand the theory of relativity. … General Relativity has become a respectable subject [so] that it is almost impious to show forth its limitations, and anyone who thinks for himself is a heretic.[10]

To Svein Rosseland,

> My little book comes out … as a lamb to the slaughter. It amounts to about 360 pages and looks devilishly long. But it is as simple (I hope) as a Bible story.[11]

The ludicrous nonsense of pretending it was like a Bible story betrayed Milne's sensitivity to a fuss about his chapter on creation. In it he traced 'the finger of God in the divine act of creation',[12] an event he placed outside human experience. Omitting the chapter, he felt, would be 'to flinch before important issues'.[13]

The book belonged to an international series of monographs, and companion volumes such as Dirac's great classic *The Principles of Quantum Mechanics* did not dabble in metaphysics. The editors were Ralph Fowler and the Russian physicist Peter Kapitsa,[14] who, for the past thirteen years,

---

[8]Letter from Milne to G. Milne, 9 December 1934, EAM, MS.Eng.misc.b.423.

[9]Letter from Milne to H.N. Russell, 6 February 1934, Russell papers, Princeton University Library.

[10]Letter from Milne to S. Chandrasekhar, 27 September 1934, EAM, MS.Eng.misc.b.423.

[11]Letter from Milne to S. Rosseland, 19 March 1935, SR, RA/PA-1514.

[12]Milne, E.A. *Relativity, Gravitation and World Structure*, Oxford: Clarendon Press, 1935, p. 134.

[13]Ibid. p. 140.

[14]Pyotr Leonidovitch Kapitsa (1894–1984). Director of the Mond Laboratory, Cambridge from 1930 to 1934. In the USSR he was Director of the Institute for Physical Problems. Nobel Laureate 1978.

had worked at the Cavendish Laboratory. Every summer he made a trip home and came back to England, but in 1934 he was detained in the USSR against his will, never to return to Cambridge. Fowler entertained serious doubts whether Kapitsa would approve of Milne's contentious passages on God and as he was unable to confer with him in person, sent him a letter. Kapitsa replied by telegram, 'Accept Milne even with God.'[15]

Milne's insistence on placing religion within the domain of cosmology needs to be set in context. In the 1930s Britain was a church-going society and attached greater store to spiritual values than is the case today. Moreover, the provenance of other cosmological theories lay with people whose Christian convictions were no less strong. Eddington was a Quaker, Lemaître a Jesuit priest, and Edmund Whittaker,[16] who ever sought to correlate theology and physics, converted to Catholicism after weighing up different denominations of Christianity. Yet to non-believers the concept of divine creation was a stumbling block that diminished Milne's scientific credentials. On the other hand, the Christian component of Milne's and other people's cosmology was a factor which swayed Pope Pius XII to support a big bang theory of creation.[17]

Milne's jittery apprehension that his book might be ignored was groundless and there were nearly a dozen reviews of varying opinions. *The Times Literary Supplement*[18] rated the book one of the most important of the decade in physics and astronomy, dwelt on its appeal to theologians and quoted from the 'Creation' chapter. The Reverend Martin Davidson, an astronomer, enjoined theologians not to be too exultant over the birth of Milne's Universe and his avoidance of a heat-death.[19] With perceptive

---

[15]Conversation with A.G. Walker, 21 February 1990.

[16]Sir Edmund Taylor Whittaker (1873–1956). Astronomer Royal for Ireland from 1906 until 1911, when he became Professor of Mathematics at Edinburgh University.

[17]The Pope's address to the Pontifical Academy of Sciences in *Bulletin of the Atomic Scientists* 8 (1952), pp. 142–146. See Kragh, H. *Cosmology and Controversy*, Princeton: Princeton University Press, 1996, pp. 251–259, section titled 'Religion, Politics and the Universe'.

[18]'A New World Theory' in *The Times Literary Supplement* (11 April 1935).

[19]*J. Br. Astron. Assoc.* 45 (1935), pp. 295–299.

foresight McCrea judged that the significance of Milne's work would lie in its method rather than its substance.[20]

Most British reviewers disapproved of Milne's flamboyant passages on metaphysics and the way he sandwiched them between sections of mathematics. Americans, who tended to be less well versed in the philosophical aspects of science, disregarded them as irrelevant. Physicists were uneasy at his total reliance on deduction, while astronomers, regardless of whether or not they agreed with kinematic relativity, admired his extraordinary capacity to squeeze so much from his simple postulates.[21]

Eddington was withering. Ever keen on rigid scales, he could not forbear from mocking Milne's use of light-signals to measure distance:

> When I visit the Cavendish Laboratory I do not find its occupants engaged in flashing light-signals at each other, but I find practically everybody employing rigid scales or their equivalent.[22]

Was Eddington unaware that Britain was on the brink of installing radar in her ships following its development by Lindemann and Henry Tizard?[23] Milne did not complain about Eddington's review because Eddington had courteously forewarned him:

> Many thanks for sending me your book.
>
> I, wisely or unwisely, took on the job of reviewing it for *Nature*. ... I should have preferred to continue to keep out of conflict, but I am not sure whether ... it is not more irritating to an author to have his work received in silence, than to have the feeling of objection openly stated. ...
>
> I realise that the review can scarcely be pleasing to you; but I hope you will recognise that it might have been worse if ... I had let myself go without out regard to our friendship. ...

---

[20] *The Mathematical Gazette* 19 (1935), pp. 299–301.

[21] For example E.T. Whittaker in *The Observatory* 58 (June 1935), pp. 179–188, and H.T. Davis in *Isis* 36 (1936), pp. 215–218.

[22] 'Relativity and Cosmogony' in *Nature* (27 April 1935), pp. 635–636.

[23] Sir Henry Thomas Tizard (1885–1959). Chemist. Rector of Imperial College London from 1929 to 1942. Chief Scientific Advisor to the Ministry of Defence from 1948 to 1952.

I am just starting on an 8-day bicycle tour tomorrow and letting the worries of the universe fade into their proper insignificance.[24]

Predictably, Howard Robertson's unfavourable notice vexed Milne. Robertson had already taken him to task[25] for not basing his theory on empirical results and one of Robertson's papers seemed to appropriate the gist of Milne's ideas without acknowledgement. Subsequent investigation[26] shows that he wrote it before Milne's book came out although it was published afterwards. Milne felt that George McVittie[27] was 'unfair'[28] because he picked piecemeal at ideas and misunderstood them, as did Robertson and also Herbert Dingle,[29] ever a severe critic, whose long campaign against special relativity would keep him at the forefront of controversy for years to come.

As soon as the book appeared Milne's mental energy resurfaced with tremendous vigour. Fired with a fresh sense of purpose, he ambitiously set himself the task of reframing the scope of his work. Kinematic relativity was no longer to be concerned just with the structure of the Universe. It was to be the springboard for reconstructing the very foundations of mathematical physics, based on the observer's consciousness of the passage of time and executed with the rigour of a geometric proof. He drew inspiration from the spiritual interpretation that Bishop Barnes gave to science in

---

[24]Letter from A.S. Eddington to Milne, 8 April 1935, in the author's possession.

[25]Robertson, H.P. 'On E.A. Milne's Theory of World Structure' in *Zeitschrift f. Astrophysik* 7 (1933), pp.153–166, supported, in the same volume, by Dingle, H. 'On E.A. Milne's Theory of World Structure and the Expansion of the Universe', pp. 167–179, and Milne's reply 'Note on H.P. Robertson's Paper on World Structure', pp. 180–187.

[26]Urani, J. and Gale, G. 'E.A. Milne and the Origins of Modern Cosmology' in Earman, J., Janssen, M. and Norton, J.D. (eds) *The Attraction of Gravitation*, Springer: Birkhäuser, 1993, p. 416.

[27]George Cunliffe McVittie (1904–1988). Professor at Queen Mary College, London from 1948 to 1952, and at the University of Illinois, Urbana from 1952 to 1972. His review is in *Philosophical Magazine* 20 (1935), p. 1068.

[28]Letter from Milne to S. Chandrasekhar, 16 November 1935, EAM, MS.Eng.misc.b.427.

[29]Herbert Dingle (1890–1978). Professor at Imperial College London from 1937 to 1946 and at University College London from 1946 to 1955. Founded the British Society for the Philosophy of Science and the *British Journal of the Philosophy of Science*. His review is in *Philosophy* 11 (1936), pp. 95–97.

his Gifford Lectures,[30] which chimed with Milne's conviction that the laws of nature governing the physical world are the expression of God's will.[31]

The audacity of Milne's scheme[32] was only matched by the remarkable degree to which he achieved it.[33] Within weeks, to his astonishment, he was close to deriving Newton's laws of motion.[34] Milne found his dynamics 'exciting beyond belief'[35] and joked that his method of arriving at the inverse square law of gravitation 'would have interested old Newton'![36] Milne's electrodynamics produced 'most beautiful'[37] formulae. Bumptiously he told his brother,

> I was visited by a sudden flash of inspiration [and] after a few days [of] intensive and wildly exciting mathematics I got out the complete structure of dynamics from *a priori* considerations … It goes amazingly deep and I think it ranks in importance with the dynamical sequence [of] Galileo, Huygens, Newton and Einstein.[38]

In full spate he also produced more papers on stellar structure.

He was riding high on other fronts, too. At this time he was vice-president of both the Royal Astronomical Society and the London Mathematical Society, and would progress to president.[39]

In July he took Margot to Paris for meetings of the International Union of Astronomy. They stayed on the Rive Gauche, within browsing distance of his favourite mathematical bookshop, Gauthier-Villars, on the Quai des Grands Augustins. He made light of the commissions he served on and

---

[30] Barnes, E.W. *Scientific Theory and Religion*, Cambridge: Cambridge University Press, 1933.

[31] Milne, E.A. 'Dr. Barnes as a Scientist' in *The Oxford Magazine* 51 (1933), pp. 730–731.

[32] Summarised by Milne in his presidential address to the London Mathematical Society, *Journal of the London Mathematical Society* 15 (1940), pp. 44–80.

[33] Nine papers in *Proc. Roy. Soc.* between 1936 and 1938, five papers in *Phil. Mag.* in 1943.

[34] Letter from Milne to S. Chandrasekhar, 30 June 1935, EAM, MS.Eng.misc.b.427.

[35] Letter from Milne to S. Chandrasekhar, 11 October 1935, EAM, MS.Eng.misc.b.427.

[36] Fragment of letter from Milne to G. Milne, 10 April 1936, EAM, MS.Eng.misc.b.423.

[37] Letter from Milne to A.G. Walker, 22 June 1937, AGW.

[38] Letter from Milne to G. Milne, 28 July 1935, EAM, MS.Eng.misc.b.423.

[39] President of the London Mathematical Society from 1937 to 1939. President of the Royal Astronomical Society from 1943 to 1945.

preferred to socialise with foreign astronomers in a programme not without splendour.

> I met again Germans, Swedes, Danes, Norwegians and Dutch of my acquaintance, and some new Frenchmen. We were received by the President of the Republic, and entertained to a garden party at the Elysée [Palace] ... climbed up the Eiffel Tower and saw Paris flood-lit, and visited Fontainebleau and Versailles ... Champagne flowed like water, but there were some rare scrambles for refreshments, owing to inadequate organisation — plenty of food and too few waiters.[40]

These middle 1930s were intensely satisfying and fulfilling for Milne. He had established himself within the university, his work was stimulating and he was blessed with a happy home life. Margot's visitors' book shows that among their overnight guests were Patrick Blackett, Chandra, Ralph Fowler, Douglas Hartree, Louis Mordell, H.W. Richmond, Henry Norris Russell and Geoffrey Taylor, besides various Milnes and Fiddeses. One weekend Philip helped them plant some two hundred spring bulbs. When Geoffrey had extended leave in England to construct his definitive soil map of East Africa, he and Kathleen left their one-year-old son in Margot's care for a week.

In the early summer of 1936 Megnad Saha and his wife spent a few months in England and before making Oxford their centre consulted Margot about accommodation and hiring a maid. Milne gently warned Saha that although Lindemann was always polite he might be touchy, because he had 'never forgiven himself for not discovering the physical nature of the stellar spectra sequence himself, which we all recognise was due to you'.[41]

That July, aboard a ship to Oslo for the International Congress of Mathematicians, Milne came across the mathematician John Synge.[42] Although Synge broadly agreed with the supremacy of time over space,

---

[40]Letter from Milne to G. Milne, 28 July 1935, EAM, MS. Eng.misc.b.423.
[41]My thanks to David DeVorkin for this extract from Milne's letter to M.N. Saha, 7 September 1935, Nehru Memorial Library, New Delhi, where Saha's diary is held.
[42]John Lighton Synge (1897–1995). Mathematician. He held appointments in Canada, America and Dublin.

he had mounted a vigorous attack on Milne after reading his book. So Milne was pleasantly surprised by the 'Irishman from Dublin, young-looking, and very gay. I enjoyed seeing him.' He said he was merely 'trailing his shirt in writing to *Nature*'.[43] Like his uncle, the playwright John Millington Synge, he had a gift for words, and used it to good effect by increasing the public understanding of science.

During the period that kinematic relativity dominated cosmology it inspired about seventy papers.[44] Of permanent value were those from the Oxford mathematician Geoffrey Walker,[45] a man slight in build, self-effacing and softly spoken. Although he studied for his PhD in Edinburgh under Edmund Whittaker, he simultaneously held Merton's Harmsworth Scholarship. Milne admired his acute mind and his proficiency in algebra, which he knew exceeded his own. More than that, he put great value on Walker's judgement and liked him to cast a critical eye over his papers and articles.

Walker came to meals, played the odd game of squash with Milne and joined my parents on cycling excursions. As it happened, Walker was on friendly terms with our vicar, the Reverend George Foster Carter,[46] previously Rector of St Aldate's, who lived a couple of doors away in Northmoor Road. Just once during discussions, which ranged over all manner of things including theology, Walker silenced Milne's rapid flow of words by asserting that God belongs to a class of items that we do not understand. Milne paused, scratched the back of his head, a characteristically impulsive reaction when he was puzzled, then nodded his assent and resumed the conversation.[47] Walker kept a photograph of Milne on his desk and preserved all his postcards and letters.

---

[43]Letter from Milne to G.J. Whitrow, 14 August 1936, GJW, private access granted by GRW, contact Imperial College Archives.

[44]Urani, J. and Gale, G. 'E.A. Milne and the Origins of Modern Cosmology' in Earman, J., Janssen, M. and Norton, J.D. (eds) *The Attraction of Gravitation*, Springer: Birkhäuser, 1993, p. 400.

[45]Arthur Geoffrey Walker (1909–2001). Geometer. Professor at Sheffield from 1947 to 1952 and at Liverpool from 1952 until 1974.

[46]George Foster Carter (1875–1966). Vicar of St Andrews Church, Oxford, from 1938 to 1954.

[47]Conversation with A.G. Walker, 21 February 1990.

In his Oxford finals Walker had attained a distinction in differential geometry, and Milne suggested that he might like to investigate the orientation of the Galaxy with regard to different geometries. This appealed to Walker — in 1948 he titled his inaugural lecture at Sheffield 'Geometry and the Universe.'

Every cosmological model is intrinsically bound up with a specific type of geometry defined by its associated equations, the metric. Besides shape, the metric concerns size, because size influences the properties of curved spaces, a feature which distinguishes them from flat space with zero curvature. The question of curvature — positive, zero or negative — was the focus of much attention, and even today its nature is unresolved. The barely discernable tiny deviation from flat space is tricky to measure, nor is flat space entirely ruled out.

Milne felt that Walker did 'great things with my gravitational theory'[48] by showing that kinematic relativity is as valid for geometries with positive and negative curvatures as for flat Euclidean geometry. This led Walker to formulate the famous metric which bears his name, the Robertson–Walker metric, which defines a homogeneous isotropic model of the Universe, and which Howard Robertson enunciated independently at much the same time.

Milne's closest collaborator was Gerald Whitrow,[49] whose deep understanding of the complexities of kinematic relativity enabled him to tighten its framework and put it on a stricter footing. He became an authority on time and on the history and philosophy of cosmology and his books are translated into European and Asian languages. A big man with a dome-shaped head and a prodigious memory, his booming voice could be heard on Radio 3 after World War II.

Improbable as it sounds, it was Morris dancing that was responsible for taking Whitrow to Oxford rather than Cambridge as he had originally intended. Coming from a modest background, he won a place at Christ's Hospital, Horsham, founded in Tudor times for the poor, whose pupils

---

[48]Letter from Milne to S. Chandrasekhar, 10 March 1934, EAM.
[49]Gerald James Whitrow (1912–2000). Lecturer, reader and professor at Imperial College London. Founding President of the British Society for the History of Mathematics. First President of the International Society for the Study of Time.

proudly wear their distinctive yellow socks and navy tunics. The headmaster's love of Morris dancing brought him into contact with Theo Chaundy, which led to Chaundy becoming the school's external moderator in mathematics, a position now obsolete. He spotted Whitrow's exceptional ability and, ever on the lookout for top-flight talent, struck a deal. If Whitrow sat for a scholarship at Chaundy's college, Christ Church, instead of Trinity College, Cambridge, Chaundy would guarantee his success. Far from tempting Whitrow to slacken his pace, this proposal had the opposite effect of spurring him to extra effort.[50]

Whitrow was one of Chaundy's finest pupils. He carried off the Junior and Senior Mathematical Scholarships, as well as a coveted University Senior Studentship. He intended to do a doctorate in pure mathematics but Chaundy steered him towards Milne as offering more stimulation. As it happened, Whitrow had already written a paper on kinematic relativity as a result of attending Milne's lectures. Whitrow was not under financial pressure to rush his doctorate, for, besides succeeding Walker in the Harmsworth Scholarship, he became a lecturer at Christ Church at the young age of twenty-four. In 1939 when he submitted material for his DPhil, the bulk of it was a clutch of published papers, not a thesis.

At the heart of Milne and Whitrow's collaboration was their searching enquiry into the elusive nature of time. It seems probable that Milne's overriding fascination with it began with the Trinity philosopher, John McTaggart, who argued that time is an illusion. The plots of classics, such as *The Time Machine* by H.G. Wells and *The Picture of Dorian Gray* by Oscar Wilde, exploit tricks with time. When John Dunne,[51] an Irish aircraft designer turned philosopher, published his provocative *An Experiment with Time* in 1927, it was widely read. He cited precognitive dreams as a manifestation of our simultaneous awareness of past, present and future events. His ideas influenced[52] T.S. Eliot, Graham Greene, Aldous Huxley and J.B. Priestley, who rated Dunne 'one of the boldest and most original

---

[50]Conversation with G.J. Whitrow, 5 October 1995.
[51]John William Dunne (1875–1949), designed and flew early monoplanes and biplanes.
[52]I am obliged to Ernst Sondheimer for pointing out Dunne's influence on contemporaneous writers.

thinkers of this age'.[53] Dunne sent copies of his books to Milne and Milne took him along to an RAS meeting and club dinner.

Time is fleeting. Once gone, it cannot be recaptured for future use. As Eddington remarked, 'Nature has made our gears in such a way that we can never get into reverse.'[54] It is impossible to re-measure a time interval against a 'standard', in the way that a length can be re-checked against a standard metre. Since we cannot compare time intervals, we have no evidence that time flows at a uniform rate. Perhaps it fluctuates, sometimes flowing faster, sometimes slower. How might this affect so-called 'constants' of nature? What do timekeeping devices tell us about the intrinsic nature of time? These questions intrigued Milne and the answers were highly relevant to kinematic relativity.

The first step in setting up a timekeeping system is to define an instrument for keeping time, i.e. a 'clock'. This, according to Milne, 'is simply a device for making a chronological record in a chronological language'.[55] The next step is to synchronise clocks between observers (on different galaxies), by exchanging light signals. These 'equivalent' observers can exchange 'clocks' with impunity since their clocks are congruent, in perfect symmetry. In creating this common time, complications that relate to the relative motions of the observers give rise to theorems in dynamics. It was for these theorems that Milne cherished the hope of showing that they were compatible with empirical physics.

Milne pondered whether the laws of physics themselves — the laws of nature, he called them — might vary over long periods of time. Just because they appear to have remained constant over recent centuries is no proof that they have always behaved like this or that they always would in the future. He supported Mach's principle which roughly states that the behaviour of physical bodies is influenced by the mass of the totality of all objects within the Universe. As the Universe expands, the distribution

---

[53] Words from J.B. Priestley's review in the *Spectator* of *Nothing Dies* by J.W. Dunne, quoted on the dust jacket of Johnson, M.C. *Time, Knowledge and the Nebulae*, London: Faber and Faber Ltd, 1945.

[54] Eddington's presidential address to the Mathematical Association, *The Mathematical Gazette* 15 (1930–1931), p. 316.

[55] Letter from Milne to G.J. Whitrow, 23 September 1936, GJW, private access granted by GRW, contact Imperial College Archives.

of matter alters. The resulting decrease in density reduces the strength of gravitational attraction and might lead to a modification in the laws of physics.

In common with most big bang theories, Milne's assumed the existence of a moment when the expansion began, i.e. the birth of the Universe. This in turn implies that the Universe has attained a certain age. The question as to what that age might be was the subject of a lot of speculation. A straightforward answer seemed to follow from the rate of expansion. The Hubble Constant quantifies the proportionality between the speed of recession of the galaxies and their distances, so if we assume the expansion is regular, an imaginary 'winding back' in history gives an estimate of how long ago the expansion started. By this reckoning the age of the Universe was about 2 billion years (2×10^9 years).

Many people wanted to accept the calculation, but it threw up an insuperable snag, a glaring contradiction. The figure of 2 billion years was far less than the age of the Earth, the Solar System and even the stars. Measurements of radioactive decay in rocks indicated that the age of the Earth was about 3 or 4 billion years and there was reliable evidence that the stars were about 10 billion years old. Clearly it was impossible for the Universe to be younger than its constituent parts! Here was a pretty paradox of Gilbertian proportions. It influenced the atheist Fred Hoyle, who disagreed with cosmology that included creation, and led him to propose his 'steady state' theory of an unchanging Universe with an infinite past.

For the next two decades the paradox was the bane of cosmologists. (A comparable modern puzzle is the whereabouts of the vast quantity of invisible dark matter in the Universe. Where is it lurking?) In 1952, after Milne's death, Walter Baade[56] announced a dramatic revision of the age discrepancy. In the wartime blackout of Los Angeles which made the night sky darker, he was able to discern from Mount Wilson a new class of Cepheids that have a different luminosity–period relationship. He distinguished stars in the spiral arms (Population I) of the Andromeda Galaxy from older stars in its centre (Population II). Calculations using the new galactic distances doubled the estimate for the age of the Universe. Nowadays, with refined

---

[56]Wilhelm Heinrich Walter Baade (1893–1960). German-born astronomer. He left Hamburg for the USA in 1931. During World War II he was placed under curfew.

measurements and spacecraft technology, the age is usually reckoned to be about 13.7 billion years, while the age of older stars is no more than 12 billion years. Thus the conundrum, as Milne knew it, no longer exists.

Although he was disturbed by the age paradox, he did not deliberately set out to untangle it, but it vanished by sleight of hand when he rejigged assumptions about timekeeping. He made the startling proposal that two kinds of time exist, each with its own clock and scale. The two scales, he explained, were like two languages for describing the Universe: kinematic time (t) applies to atoms and stars in an expanding Universe; dynamical time (T) applies to pendulums and systems obeying Newtonian mechanics, such as the motion of the planets around the Sun. (In fact he believed many time scales existed, but selected two for their physical applications.) Later he went even further by suggesting that the expanding Universe itself is a clock.[57]

A simple formula connects the scales:

$$T = t_0 \log(t/t_0) + t_0$$

where t_0 is the present age of the Universe. When the Universe began, t takes the value zero, and on the T scale this special state of affairs defies description; it is a singularity. This is analogous to zero on the Kelvin temperature scale when molecular motion stops and matter ceases to exist. Even the laws of physics have no meaning. A by-product of the scheme was the disappearance of the age paradox: on the t scale the Universe has a definite beginning and a finite age, but on the T scale the world is time-less and infinitely old.

In September 1936 Milne released his weird idea at the British Association for the Advancement of Science in Blackpool, where among those present was H.G. Wells.[58] Milne chose a moment during a discussion on the origin of the Solar System to make his sensational announcement. It prompted the chairman of the session to accuse him of 'throwing a brick into the machine with a vengeance'.[59] Milne told a newspaper reporter that we have no evidence that atoms and pendulums keep the

[57]Whitrow, G.J. *The Nature of Time*, London: Pelican, 1972, p. 125.

[58]Crowther, J.G. *Fifty Years with Science*, London: Barrie & Jenkins, 1970, p. 168.

[59]'Origin of Solar System' in *The Times* (11 September 1936).

same sort of time, and that it is no harder to envisage two sorts of time than to treat time as a fourth dimension.

Jeans thought Milne was on to 'something very big'[60] and offered a helpful metaphor. That Milne painted two different canvases, each giving valid pictures of the Universe, was akin to mapping the Earth's surface in different projections. (Mercator preserves shape but distorts area, whereas other projections preserve area and deform shape.)[61] Furthermore, a sufficiently small area is the same on all maps regardless of projection. Similarly, the two time scales coincide over short intervals of time, such as the present epoch. The left-wing polymath Desmond Bernal,[62] who unravelled the crystalline structure of proteins, added the metaphor of sound. Octaves extend upwards and downwards to infinity; yet, if the frequency of a note is measured in vibrations per second this can reduce to nil vibrations per second.[63]

Milne's proposal had a conspicuous impact. By enriching the ambit of kinematic relativity, it consolidated his central position in cosmology and made him the predominant force in its development.[64] To date, the two kinds of clock have not found an application, but, as the philosopher of science Mary Hesse[65] (whom Milne examined for her doctorate) points out, the nature of time continues to be enigmatic, and Milne's ideas should not be overlooked.[66]

The two time scales had implications for the 'constants' of physics, whose values are crucial in shaping the nature of our Universe. Milne floated the idea that the value of Planck's constant h and the value of the gravitational constant G vary over long periods of time. On his T scale, G is constant while on the t scale it gradually increases with time, provided that the mass of the Universe remains unchanged.

---

[60]'New Theory of Time' in *The Morning Post* (11 September 1936).

[61]Jeans, J.H. *The Universe Around Us*, 4th ed., Cambridge: Cambridge University Press, 1944, pp. 92–94.

[62]John Desmond Bernal (1901–1971). Professor at Birkbeck College, London from 1938 to 1968.

[63]'The Birth of the Planets' in *The Daily Worker* (10 April 1945).

[64]Kragh, H. 'Cosmo-physics in the Thirties' in *HSPS* 13 (1982), p. 77, and Kragh, H. *Quantum Generation*, Princeton: Princeton University Press, 1999, p. 226.

[65]Mary Brenda Hesse (b. 1924). Professor of the Philosophy of Science at Cambridge from 1975 to 1990.

[66]Conversation with Mary Hesse, 29 October 1992.

Paul Dirac took up the theme of the two time scales in his own cosmology,[67] though his investigations pointed to the opposite result, that G decreases with time.[68] Like Milne he believed in the paramount power of mathematics, and declared that 'Mathematics can lead us in a direction we would not take if we only followed up physical ideas.'[69]

As the number of cosmological theories proliferated, so the scrutiny of their epistemological and philosophical merits intensified. The twentieth century threw up a large choice of models, starting with Friedmann in 1922, then Einstein, De Sitter, Lemaître, Whittaker, Eddington, Milne, Dirac and Jordan.[70]

Milne and Eddington may have differed over stellar structure but they shared a passionate conviction about the infallibility of their mathematical methods, which nicely illustrated G.H. Hardy's dictum that mathematics creates very strong philosophical prejudices.[71] By flaunting their obsession with pure reasoning they made themselves, along with Dirac, the target of a vitriolic tirade from Herbert Dingle.

Under the title 'Modern Aristotelianism', Dingle berated theorists like Aristotle, for emphasising ideas rather than facts and for applying (syllogistic) logic to generate a new statement from two linked premises, by contrast to Galileo, who devised theory to fit observations. Dingle lambasted the notion that laws of nature could be derived without recourse to experiment, and condemned the publication of 'spineless rhetoric ... obscured by a smoke screen of mathematical symbols'.[72] In a recent article in *Philosophy*[73] he had poured scorn on Milne and now reserved his

---

[67]Kragh, H. *Dirac: A Scientific Biography*, Cambridge: Cambridge University Press, 1990, pp. 229, 240.

[68]Dirac's letter in *Nature* 139 (5 February 1937), p. 323.

[69]Quoted by Marcus du Sautoy, 'Equation that Changed the World — Mind and Matter' in *The Times* (13 November 1995).

[70]Ernst Pascual Jordan (1902–1980), German quantum theorist, held appointments at Rostock, Berlin and Hamburg.

[71]Hardy, G.H. 'Mathematical Proof' in *Mind* 38 (1928), p. 3.

[72]*Nature* 139 (8 May 1937), p. 784.

[73]'On the Principles Underlying Professor Milne's Cosmological Theory' in *Philosophy* 11 (1936), pp. 48–59. The same volume carries Dingle's caustic review of Milne's book, pp. 95–97.

sharpest barbs for him. Dingle is the only person I remember my father speaking of with undisguised irritation.

The heated controversy was on a par with the famous seventeenth century argument between Locke and Descartes,[74] and ranged beyond the foundations of physics. *Nature* brought out a sixteen-page supplement[75] to air the landscape of opinions. Milne led for the defence. No one was more astonished than he was, he reminded his readers, at the results he obtained from his simple premises. Eddington declared that he did not object to the label 'Aristotelian'. Dirac wanted to see a balance between observations and deductions from speculative assumptions. Charles G. Darwin reproached Dingle for hampering speculation: 'It is hard enough to make discoveries in science without having to obey arbitrary rules in doing so; in discovering the laws of nature foul means are perfectly fair.'[76]

Although Harold Jeffreys,[77] the father of geophysics, concurred with Dingle's disapproval of purely deductive methods, he sounded out the eminent statistician Ronald Fisher.[78] Fisher replied cautiously:

> I have been a great deal interested in Milne's recent writing, and have been a little puzzled by, I think, the same point as you, namely his taste for developing the subject as much *a priori* as possible. ... I do think that his dynamical work is a marvellous achievement, and it seems at present to be in the way of clearing up the whole tangle for which Einstein and his followers are responsible.[79]

---

[74] Gale, G. 'Philosophical Aspects of the Origin of Modern Cosmology' in Hetherington, N. (ed.) *Encyclopaedia of Cosmology*, New York: Garland, 1993, pp. 487, 491.

[75] Supplement to *Nature* 139 (12 June 1937), pp. 997–1012.

[76] Supplement to *Nature* 139 (12 June 1937), p. 1008.

[77] Sir Harold Jeffreys (1891–1989). Plumian Professor at Cambridge from 1946 to 1958.

[78] Sir Ronald Aylmer Fisher (1890–1962). Appointed professor at University College London in 1933, and at Cambridge in 1943. His practical handbook *Statistical Methods for Research Workers* went into fourteen editions.

[79] I am obliged to Maurice Bartlett for telling me about the letters between Jeffreys and Fisher, in Bennett, J.H. (ed.) *Statistical Inference and Analysis*, Oxford: Oxford University Press, 1990, pp. 162–163.

J.B.S. Haldane stressed that Dingle's stinging rant should not deter biologists and geologists from studying Milne's ideas, and urged Milne to quantify possible secular changes in density, rigidity, viscosity, surface tension and conductivity.

For ideological reasons, Haldane championed kinematic relativity on the basis that Milne's 'observers' rate equally and none have privileged positions.[80] Haldane was good at explaining scientific ideas in simple terms and in one of his weekly columns for the *Daily Worker*[81] he argued why the theory endorsed communism.

Thirty years later, long after Haldane had abandoned communism for mysticism, he was gravely ill with cancer. In a televised interview[82] which amounted to an auto-obituary, he was asked to name that for which he would wish to be remembered. He immediately gave pride of place to a letter in *The Observatory* that concerned Milne's cosmology.[83]

The time scales sparked Haldane's imagination. If the laws of nature were changing, he mused, they would extend our interpretations of the world and he discussed the biological implications with Milne.[84] In *The Marxist Philosophy and the Sciences* Haldane pointed out that the human capacity to keep track of time is not sacrosanct but depends on the temperature of the brain. For instance, a person with a fever overestimates the number of seconds in a given interval.[85] When Haldane corresponded about this book with Milne, Britain was at war and Milne could not explore the time scales further because he was immersed in defence work. He thanked Haldane for paying him the compliment of crediting him with being 'intellectually honest', yet, rather disingenuously bridled at being

---

[80]Haldane, J.B.S. *The Marxist Philosophy and the Sciences*, London: Allen and Unwin, 1938, p. 62.

[81]Reprinted in Haldane, J.B.S. *Science Advances*, London: Allen and Unwin, 1947, pp. 37–40.

[82]The interview in February 1964 was rebroadcast on BBC Radio 4 on 2 August 1990.

[83]*The Observatory* 59 (1936), pp. 228–229.

[84]Letter from Milne to G.J. Whitrow, 14 April 1938, GJW, private access granted by GRW, contact Imperial College Archives.

[85]Haldane, J.B.S. *The Marxist Philosophy and the Sciences*, London: Allen and Unwin, 1938, p. 66.

labelled a theist. He doubted whether he was always a theist, but death encouraged belief.[86]

Death was never far from Milne's mind because he was now a sad widower. After the birth of my brother in June 1938 my mother suffered severe post-natal depression. She was admitted to hospital, my brother went back to the maternity home and my father cancelled the summer holiday on a Somerset farm, planned with the new baby in mind. Struggling to cope with domestic cares, he cobbled together makeshift arrangements while urgently writing, telephoning and telegraphing in all directions to find a housekeeper. On top of this, he was required to do jury service.

It must have been a terrible time for my father. He was now in charge of three small children and beside himself with anxiety as my mother's health fluctuated. Weeks went by and money worries began to loom as fees for the hospital and maternity home mounted. By August my father was overcome with exhaustion and utterly demoralised, fearful that his years of mathematical creativity were over. He unburdened himself to McCrea.

> I realise that I am at best a very second rate mathematician, whereas I always hoped to turn myself into a good one. ... I am in fact an ignoramus. Men of science like yourself become increasingly mature, whilst I get more elementary.[87]

Like a true friend McCrea bolstered my father to regain a 'firmness of mind'[88] but on 5 October 1938 the desperate situation ended grimly. The vigilance of the hospital slipped and in an unlocked shed of garden tools my mother took her life. My father was devastated and I don't think he ever recovered from the shadow of her death. He used to console himself with the thought that one day he would join her in heaven.

Distraught, forlorn, he had to face the harrowing ordeal of a public inquest. Suicide was shameful, a taboo subject not to be mentioned and a crime until the law was changed in 1961. As for the distasteful question

---

[86]Letter from Milne to J.B.S. Haldane, 10 November 1939, Haldane papers, University College London Library Services, Special Collections.
[87]Letter from Milne to W.H. McCrea, 12 August 1938, WHMcC.
[88]Letter from Milne to W.H. McCrea, 16 September 1938, WHMcC.

of whether the hospital was culpably negligent, my father consulted my mother's cousin, David Maxwell Fyfe,[89] the son of Hugh Campbell's sister, a formidable lawyer, clear-headed and extraordinarily industrious. To qualify at the bar without money or connections was a great feat in those days and when he took silk at the age of thirty-four he was an exceptionally young King's Counsel. He entered parliament and rose to be Lord Chancellor. At the post-war Nuremberg trials his experience of prosecuting convinced him of the need for a European Convention of Human Rights and he played a considerable part in drafting it.

He advised Milne against seeking redress. Litigation would add to his emotional strain and of course nothing could bring my mother back. Friends rallied with practical help. On the day of my mother's burial at Wolvercote Cemetery, the Robinsons looked after my sister and me in their big house at 117 Banbury Road. The Oxford mathematician Ted Titchmarsh, himself the father of three daughters, solved our domestic difficulties. His sister Kathleen Titchmarsh left her job at a boys' preparatory school and became our trusty housekeeper.

It is hard to imagine what my father must have gone through in the dark months that followed. Numb with grief, he lost all incentive to engage in mathematics yet managed to keep up an outward semblance of discharging his duties, lecturing, attending meetings and holding his colloquia. Inwardly his brain was sterile, 'stiff and set'[90] and his research ceased.

---

[89] David Patrick Maxwell Fyfe, Lord Kilmuir of Dornoch (1900–1967). Solicitor General, Attorney General, Home Secretary, and Lord Chancellor from 1954 to 1963.
[90] Letter from Milne to A.G. Walker, 20 January 1939, AGW.

Chapter 14
Lifeline

By taking a second wife he pays the highest compliment to the first, by showing that she made him so happy as a married man.[1]

Two friends hoisted Milne out of his abyss of sad inertia. Chandra judged that a change of scene might revive his spirits and rekindle his love of mathematics. A perfect pretext was the up-coming gala inauguration of the new McDonald Observatory in Texas and he approached Otto Struve,[2] the director of both the fledgling observatory and of Yerkes Observatory, who agreed to invite Milne to the celebrations. That the new observatory had ever reached completion was largely due to Struve. He seized the initiative when the project foundered despite an enormous bequest from William Johnson McDonald, banker and amateur astronomer.

Struve added an extra inducement for Milne by asking him to contribute to a symposium. Chandra wrote pressing Milne to spend a few days with him and his wife Lalitha in Chicago, and simultaneously Milne received the formal announcement of the opening of the observatory from the President of Warner and Swasey, the company that made the precision instruments for the observatory and that spared no expense in hosting the lavish entertainment. Under this barrage, Milne gratefully cabled his acceptance. Besides providing a sorely needed distraction from his sorrow, the invitation to speak at the symposium motivated him to restart his research.

---

[1]Milne marked this sentence in his copy of Boswell's *Life of Johnson*, London: Henry Frowde, 1904, p. 384.

[2]Otto Struve (1897–1963). In 1921 he left Russia for the USA. Director of Yerkes Observatory from 1932 to 1947, of McDonald Observatory from 1939 to 1950 and of the National Radio Astronomy Observatory of the University of Virginia from 1952 to 1962. His father, grandfather and great-grandfather were eminent astronomers.

Oxford University granted Milne a month's leave of absence, most of it absorbed in travel, and on 22 April 1939 he sailed for New York. After monotonous days at sea, Manhattan's distinctive silhouette rising over the horizon is a splendid sight, and once ashore Milne explored its famous buildings. In Chicago he joined the large party of astronomers on the deluxe Pullman train with sleeping-cars chartered by Warner and Swasey. After a 2,000 mile (3,200 km) journey the astronomers alighted at Alpine, the nearest railhead, in a violent thunderstorm that pelted them with huge hailstones.

The final forty miles across ranch country and up into the Davis Mountains took the party to the rocky peak of Mount Locke, 2,100 metres high, where the observatory enjoys viewing conditions of unrivalled clarity. The great gathering of astronomers was the last of its kind before World War II and Milne enjoyed the companionship of many friends and colleagues: Bertil Lindblad from Sweden, Jan Oort from the Netherlands, John Plaskett from Canada and Americans aplenty — Harlow Shapley, Cecilia Payne Gaposchkin and her husband Sergei from the Harvard Observatory, Walter Baade and Edwin Hubble from Mount Wilson and Henry Norris Russell from Princeton (Fig. 17). Besides astronomers, a bevy of engineers, university officials and Texas residents made up some four hundred guests.

Ceremony and speeches alternated with informal entertainments. Oxford must have seemed a world away as Milne partook of the chuck-wagon barbecue and watched cowboys handling broncos and lassoing cattle at an improvised rodeo.

Hubble chaired the symposium at which Milne posed six fundamental questions about the Universe: Is its past history finite? Is it expanding or not? Is space curved or flat? and so on, the big questions that we still ponder today. Milne claimed that while other theories gave ambiguous answers, only kinematic relativity supplied definite ones, but they varied according to the scale of time adopted. This flexible interpretation smacked of artificial contrivance and did not go down well with his listeners.

After a short visit with the Chandrasekhars, Milne headed for home. The string of experiences was a marvellous tonic and he was in fine fettle (Fig. 18). Never had he travelled so far nor seen so many new sights. He wrote a chirpy account[3] of his visit and told Theodore Dunham, 'Your great country has done much to restore my outlook on life ... I cannot tell you how re-vivifying [the trip] has been.'[4] And he had a private reason for feeling rejuvenated for his heart was set aflutter by an American woman he met aboard ship.

The 900-foot (273 metre) long RMS *Aquitania* with her four distinctive red smokestacks was one of Cunard's grandest liners; its interior was designed by the person who did the Ritz Hotel. In the bracing, salty air Milne exercised by pacing around the deck, lap after lap. As he made his circuits he kept passing a lively young woman doing laps in the opposite direction and inevitably they fell into conversation. After his recent busy social programme, now alone in the romantic ocean setting and isolated from his cares at home, he was at his most vulnerable, and he fell under the spell of her beguiling charms. Her name was Beatrice Brevoort Renwick and she was going to see her sister Izzy and English husband in Kent. During her visit she and Milne met twice — once to meet us children — and they became informally engaged. She was twenty-seven years old to his forty-three.

Beatrice was the youngest of three daughters of a distinguished 'old' New York family with Dutch and Scottish roots. The Brevoorts emigrated when the city was called New Amsterdam and bought land north of Washington Square. Legend has it that their refusal to sell their cherry orchard caused the 'kink' in Broadway. Eighteenth-century Renwicks ran a shipping service between Liverpool and New York. Nineteenth-century Renwicks excelled as inventors, engineers and architects. Beatrice's grandfather, Edward Renwick,[5] ingeniously repaired the enormous eighty foot (24 m) long gash in the hull of Brunel's *Great Eastern*, a task that had been declared impossible because the steamship was too big to fit into a dry dock. Twentieth-century Renwicks were equally independent and resourceful. They thought for themselves, shrugged off social niceties and conventions, were indifferent to fashion and did not follow the herd.

---

[3] 'Dedication of the McDonald Observatory' in *The Observatory* 62 (1939), pp. 187–191.
[4] Letter from Milne to T. Dunham, 24 May 1939, EAM, MS.Eng.misc. b.429.
[5] Edward Sabine Renwick (1823–1912). Inventor and patent expert.

On two counts Beatrice's background heightened her irresistible allure. By an extraordinary twist of fate, she had links with astronomy through her other sister who had married (and divorced) Douglas Campbell, son of Wallace Campbell,[6] director of California's Lick Observatory. Equally bizarre, given Milne's abiding interest in churches, the family firm of Renwick, Aspinwall and Renwick were leading ecclesiastic architects. Beatrice's great-uncle, James Renwick,[7] Edward's brother, was the celebrated architect of Manhattan's neo-Gothic St Patrick's Cathedral. His career had blossomed after winning, at the age of twenty-five and without formal training, a competition to design Grace Church at Broadway and Tenth Street. (By then the Brevoorts had sold the site of the cherry orchard to the Episcopal Church.) His clever cost-cutting scheme for its pillars, using brick faced with stone, avoided the expense of solid stone. Among his secular landmarks are Vassar College in Poughkeepsie, and in Washington DC the Smithsonian 'Castle' and the Corcoran Art Gallery, now the Renwick Gallery. Beatrice's father, William Whetton Renwick,[8] who added the chantry and outside pulpit to Grace Church, devised his own style of decorating altars and walls, 'fresco relief', distilled from studying sculpture in Paris and painting in Rome. His largest commission was the Cathedral of St Peter and St Paul in Indianapolis.

Beatrice was brought up in her parents' fine house in a rural part of New Jersey. Like her sisters she was taught by a governess but the girls were not cosseted. She was an 'outdoor' girl and enjoyed rugged pursuits such as trekking on horseback through the Canadian Rockies. Cosmopolitan visits to relatives in Italy and France broadened her outlook. Artistic like her mother, who painted landscapes, Beatrice patiently watched small animals in order to sketch them and carve them out of wood. When she met Milne she was working at a health centre in Manhattan while in her spare time she ran a Girl Scout troop in a disadvantaged area of the city and took the girls camping. She was strong in body and mind, had a

---

[6]William Wallace Campbell (1862–1938). Director of Lick Observatory from 1900 to 1930. President of the University of California from 1923 to 1930.

[7]James Renwick (1818–1895). Other major commissions were St Bartholomew's Church on Madison Avenue, a house for J.P. Morgan and buildings on islands in the East River.

[8]William Whetten Renwick (1864–1933). Church architect.

no-nonsense expression and a forceful, humourless and dominating personality.

Milne felt it would be rash to announce their engagement until they knew each other better and they intended that before long one of them should cross the Atlantic so that they could spend some time together. Yet for Beatrice a hidden catalyst lay behind her engagement. She was under family pressure to find a husband and settle down like her sisters. If she acquired a husband they would stop nagging, and in their maverick eyes an Oxford professor would be quite a catch. For his part, Milne carefully considered whether to remarry. Leaving aside his own happiness, he reckoned that marriage would give his children a wider variety of experiences than he could provide on his own.

If Chandra's scheme took an unforeseen twist, A.V. Hill's effort to rehabilitate Milne ran a straighter course. In January 1939 Hill gave him lunch at his club, the Athenaeum, and coaxed him into taking up the threads of his ballistics research for the Admiralty and War Office. Ever conscious of his obligation to Hill, Milne agreed to undertake this work on a part-time basis alongside his university commitments. He hoped that his absorption in abstract problems had not 'robbed [him] of the common sense which he would need'.[9]

Only in 1936 had Britain begun to look to her defences. Until then the country, horrified at the prospect of more carnage, had been in no mood to re-arm and coasted along in the giddy post-armistice years. As Europe teetered on the brink of another war, the Royal Society drummed into the government the importance of applying scientific methods to warfare and the urgent need to draw up a register of scientists and mathematicians. This would enable them to be matched to suitable problems, unlike the haphazard allocation of the previous war, which depended on personal networks. Hill and a colleague compiled the scientific section of the central register and nobody was better placed to recruit Milne.

He was to be employed by the Ordnance Board, a body set up in Elizabethan times to convene experts in artillery against the threat of invasion by the Spanish Armada. By 1939 the board was part of the Ministry of Supply and responsible for designing weapons across the armed services. Its

---

[9]Letter from Milne to A.V. Hill, 25 January 1939, AVHL I.

presidency rotated between a Major-General, a Vice-Admiral, and an Air Vice-Marshal. Personnel were largely drawn from the services and Milne was among a sprinkling of civilian mathematicians, chemists and engineers.

Somewhat to his astonishment, once Milne was immersed in the work, he enjoyed it. He renewed contact with his old haunts of Teddington, Whale Island and Woolwich and paid a visit to Biggin Hill. He noticed that ballistics was 'terribly mixed up with personalities', that people tended to take sides on problems rather than approaching them objectively and that it would be 'necessary to tread warily'.[10]

Although planes flew higher and faster than in the previous war, techniques in anti-aircraft gunnery had hardly changed and many of the Brigands' findings were still in current use. One of his first assignments was to extend Fowler, Littlewood and Lock's famous paper on a spinning shell to quantify how the addition of a tail fin might affect its motion.

Britain declared war on 3 September 1939 and instantly the framework of Milne's life changed. He was classified as a 'key scientist' and ordered to work full-time at the Ordnance Board's wartime headquarters in Chislehurst, south London. He had no choice but to commute weekly from Oxford and from Monday to Friday he lodged at 62 Court Road, Eltham, a short bus ride from Chislehurst. Clearly going to see Beatrice in America was out of the question.

Concurrently Beatrice gained the stigma of alien status, as did all citizens of countries not involved in the war. This did not thwart her desire to reach England and she applied for a visa. The British Home Office was slow to process her papers, and she was unable to take the berth she had booked for mid-May 1940, but her visa arrived in time for her to sail on 25 May aboard the Italian *Conte di Savoia* bound for Genoa. It would be the last time the liner crossed the Atlantic before Mussolini joined the Axis powers on 10 June.

Norway, Holland and Belgium had already fallen to the Nazis and although she must have known that getting to England was risky, Beatrice embarked with typical Renwick pluck on what might have been a foolhardy disaster. During her voyage Nazi troops reached Abbeville on the French coast forcing Britain's desperate evacuation from Dunkirk.

---

[10]Letter from Milne to A.V. Hill, 17 February 1939, AVHL I.

After docking Beatrice took refuge with an uncle in Florence. She did not reveal her whereabouts to Milne but he guessed where she was and cabled her to hurry to him with all speed. On 6 June she got to Paris just ahead of the advancing German army, to be trapped there for two days among the mêlée of people like herself who were frantic to escape from any port that was still open. She managed to catch a train to St Malo, where, having spent her French francs and unable to exchange Italian lira or American dollars, she went hungry and without a bed. At last she scrambled aboard a ferry to Southampton and in London she and Milne were jubilantly re-united, a year to the day since they had last seen one another.

Under severe wartime strictures, Beatrice had to register with the police at Bow Street, which entailed hours of queuing. She was not allowed to go to Oxford until Milne obtained written permission from the city's chief constable. Equipped with this, she had to queue to de-register and then, having re-registered in Oxford, she was forbidden to move again. She stayed with our good neighbour Mrs Underhill at 10 Northmoor Road and ate her meals at No. 19.

England was at crisis point. Hitler's army glowered menacingly across the Channel with invasion expected any day. Given that Beatrice was confined to Oxford and that Milne had to work in London it was not feasible to deepen their acquaintance and they resolved to marry immediately. As the wife of a British citizen she would shed her restrictions and be completely free to come and go as she pleased.

Milne did his utmost to provide a good wedding. He scurried about to obtain a marriage licence, spoke to the vicar and bought a ring. Early on Monday, before catching his train to London, he went to Wadham to arrange a small reception with champagne, ice creams and strawberries, and hire waiters. During the week he wrote invitations, placed an announcement in *The Times* and booked a hotel in Chipping Camden for their honeymoon.

On Saturday 22 June, eleven days after Beatrice touched British soil, the Reverend George Foster Carter married the couple at St Andrew's Church. The physicist Idwal Griffith gave Beatrice away, the mathematician Jack Thompson was best man and my sister and I were bridesmaids, dressed in our best but rather outgrown Liberty-print dresses. At the reception of some fifty guests, Maurice Bowra toasted the health of bride and groom. A photograph of the couple shows them glumly seated, side by side, without a vestige of affection (Fig. 19).

The hasty marriage, born of a shipboard romance, was a by-product of the exigencies of war, yet it was utterly out of character for Milne to be so precipitate. He tended to caution and was not given to impulsive acts; quite the opposite. Twice, at Manchester and at Oxford, he had postponed taking up new appointments. Undoubtedly they were fond of each other, but it is hard to discern what they had in common. Certainly Beatrice had not the slightest interest in intellectual pursuits. Some people thought she had a bewitching pull over him. In truth it was a marriage of convenience. They were little more than strangers; they had not explored each other's whims, tastes and foibles, nor known the progressively intimate tiers of courtship. It was two years before they had an evening out together, to a Russian ballet and opera.

The situation bore harder on Beatrice, cut off from friends and family. Her sister and husband had departed for America so that he would avoid conscription. Aware of Beatrice's predicament, Milne tried to do his best, yet started off with an appallingly callous blunder. Margot's memory clung to him — he wore her wedding ring on his little finger — and as soon as the wedding guests had left, he took Beatrice by taxi to Wolvercote Cemetery for her to place her bridal bouquet of sweet peas on Margot's grave. If this tactless act salved his conscience, it must have been disturbing and distasteful to Beatrice.

After four days in the Cotswolds they cut short their honeymoon. The hectic wedding had coincided with a generous invitation from Yale University, communicated by a former Rhodes Scholar, John Fulton,[11] offering hospitality for the duration of the war to children from Oxford University families. My father felt guilty that he had not put our names down. The admirable Warden of Rhodes House, Carleton Allen,[12] handled the invitation sensitively and expeditiously, and soon a group

---

[11] John Farquhar Fulton (1899–1960). Neurophysiologist. Professor of Physiology at Yale from 1930 to 1951.

[12] Sir Carleton Kemp Allen (1887–1966). Legal scholar. Warden of Rhodes House from 1931 to 1952.

of one hundred children and twenty-five mothers were ready to go.[13] To my father's consternation all places were filled but at this auspicious moment the Dunhams warmly invited Eleanor and me to come to them in California. Since the summer in Ann Arbor they had often come to England for Theodore to work with Plaskett at the OUO and to see his aunt and uncle, Carrie and Rudyard Kipling.[14] My father havered about letting us go because he regarded us as Margot's legacy to him, but Beatrice insisted and he arranged for us to travel with the Yale group. After he had accepted the Dunhams' invitation the Chandrasekhars made an equally kind offer.

In an extraordinary week my father had gained a wife and agreed to part with two of his three children. Beatrice was glad to be rid of us and one wonders whether she had sufficiently considered the irksome burden of step-children before deciding to marry. She complained about our shrill voices and resented the attention my father gave us at the weekend. She believed in exercises before breakfast, and after showing us how easily she stood on her head scorned our unsuccessful attempts. In an astonishingly candid letter warning the Dunhams of our shortcomings she betrayed her jealousy. They 'are becoming too emotionally tied up with their father and I think the best thing in the world for them at present is to be away from home'.[15]

On Monday morning it was always a wrench for my father to depart for London, and never more so than on 8 July when we left on the boat train for Liverpool. Throughout our ten days on the Atlantic in a small, old ship, the SS *Antonia,* he was sick with worry, fearful of predatory U-boats. In fact we were attacked by torpedoes but not hit.[16] As a token of gratitude for our safe arrival, he sent a guinea to our school, the Oxford High School, a gesture he repeated during our absence that kept him in touch with the headmistress.

The Dunhams were kindness itself, as were their extended families in Maine, New Hampshire, New York and Massachusetts. They gave us a

[13] Symonds, A.S. *Havens across the Sea*, An Grianan, 1992.
[14] Rudyard Kipling (1865–1936). Writer. Nobel Laureate 1907.
[15] Letter from B.R. Milne to M.P. Dunham, 16 July 1940, in the author's possession.
[16] The government stopped its evacuation programme to North America in September 1940 after the *City of Benares* was sunk and seventy-seven children drowned.

wealth of experiences, which included driving across the United States from coast to coast. At Mount Wilson, where my father never set foot, we looked down the 100-inch Hooker telescope and rode on the little mobile platform inside the dome. We saw a technician polishing the surface of the big 200-inch mirror destined for Mount Palomar. We met Edwin Hubble again; at Princeton Henry Norris Russell entertained us with origami, miraculously transforming sheets of paper into frogs and birds while carrying on a conversation with the Dunhams. Miriam Dunham told me that my father was one of the few people to match Russell in his loquacious flow of words and that Russell would stop to allow my father to have his say.

My father had a Utopian vision of our upbringing and outlined his wishes for all facets of our education — religious, academic and social. He wanted us to learn about the broad sweep of civilisation, illustrated by traditions, myths and folklore, rather than acquire isolated chunks of information. He reminded us that we were guests, and that we should be receptive to new customs: 'Very often the American way, or the French way, and sometimes the German way is better than the English way.'[17] Letters took up to a month to cross the Atlantic, every word checked by a censor who snipped out any phrase that might conceivably help the enemy. For a child, two months is an eternity and when I read my father's responses to what I had previously related they seemed stale and irrelevant. Gradually he faded into a remote figure.

Now reduced to a household of three, the Milnes could move to London. Beatrice disliked the Oxford house and thought it pretentious, so had no regrets at leaving it, but accommodation in London was hard to find because of the number of bombed buildings and the large influx of refugees. The Milnes had to make do with a two-bedroom first-floor furnished flat at 12 Glenesk Road in Eltham. The only advantage of the dreary road, some distance from the shops, was that it led directly via a footbridge to the local railway station. They moved in on 19 August to find the flat disgustingly dirty. Worse still, after dark a policeman on blackout patrol knocked on their door to reprimand them for the chinks of light seeping from their windows. Milne was vexed that the landlady had

---

[17]Letter from Milne to E. Milne, 12 April 1941, in the author's possession.

not provided adequate curtains and expected a fine. Yet away from Margot's lingering shadow, the small flat held the promise of a fresh start to their marriage, on which Beatrice could put her own stamp. She expected him to help with the chores, and instructed him how to cook, wash and iron. Milne's comfortable pre-war prosperity had vanished forever as inflation soared to twenty percent.

The flat had no telephone and Beatrice grew lonely cooped up with two-year-old Alan, to whom she was a severe step-mother. She spilled out her discontent in homesick letters to her lawyer, who took charge of her financial affairs. She pined for American vegetables — corn, squash and zucchini — but this was trivial compared to her desire to reform Milne, especially in money matters. She had no sympathy for his prudent habit of saving a portion of his salary against school fees and old age. Her affluent background blinded her to his ingrained dread of getting into debt, which was second nature to those brought up in straitened circumstances.

Chapter 15

Mathematics, Bombs and Bureaucracy

[The scientist] is driven by the same passions as other men, weakened as they are by human frailty ... and like other men, he cursed the war and was anxious to get it done with as quickly as possible.[1]

Three weeks later the Blitz began in earnest and for fifty-seven days and nights the Luftwaffe's bombers pounded London relentlessly. Civilians suffered the most. Danger was never far away and the Milnes had several close calls. Even at the best of times mathematics requires intense concentration and under these unnerving conditions Milne needed every scrap of willpower to apply his brain effectively, especially if he were short of sleep from the perilous raids or from fire-watching duty.

His office at Kemnal Manor, a Victorian building standing among oaks and chestnuts, was an obvious target and it was hit twice in a single week. Luckily Milne had stepped smartly into a corridor seconds before his windows shattered. They were patched with cardboard until the next bomb brought the ceiling down. Then they were boarded up and he was deprived of natural light. Work continued during the raids and in a colleague's office they hastily crouched behind the desk, thankful for its protection from the avalanche of glass shards that catapulted into it.

Milne kept imperturbably to his daily rhythm, but each morning as he left the flat he was tortured with uncertainty lest on his return Beatrice and Alan were gone. When they went shopping they were often obliged to scuttle into an air-raid shelter for safety, and luckily survived their parlous existence unscathed. It was too hazardous to sleep in their flat at night because tremors shook the house and the Milnes dragged their mattresses down to the ground-floor hall, where they passed the hours fitfully, disturbed by the incessant racket, the bangs of explosions and the deafening

---

[1]Clark, R.W. *The Rise of the Boffins*, London: Phoenix, 1962, p. 71.

roar of guns. One night when the horizon glowed red with fires, and bombs gutted neighbouring houses, they woke to a clatter of breaking glass. They went upstairs to find the flat ankle-deep in glass, plaster down and, most tiresomely, the lavatory smashed. But they were not yet homeless, the fate of one in six Londoners. There was no hope of immediate repairs and all they could do was tidy up and fasten sacking across the windows. After a battering like this, street sweepers collected rubble and glass by the barrow-load. Yet Milne joked that they 'were "safe, so far", as the man said who dropped from the 60th storey, when passing the 10th'.[2] Another night, the Milnes saw the house four doors down go up in flames and rushed out to help their neighbours rescue their furniture. 'It was a grim sight, fires reflected in people's faces and in tin hats, shells bursting overhead, guns flashing nearby, flames leaping to the roof.'[3]

Transport was chaotic. Bombs wiped out tracks and signals, timetables were unpredictable, and Milne had no way of telling which stations were functioning when he needed to get to meetings and gun trials. The slow, grimy trains were packed to capacity, and after dark compartments went unlit to comply with blackout regulations. For Christmas the Milnes made an exhausting journey to Hessle for Beatrice and Edith to meet. The miserable delays and the hoards of teeming people were far worse, Beatrice declared, than her escape from France, and she vowed she would not go again.

At this early stage of their marriage, especially in the frightening conditions, it would have been good if my father and his new wife could have drawn closer to each other, so it was unfortunate that he could not talk to her about his work. Having signed the Official Secrets Act his lips were sealed and his mathematical life was cordoned off from his home life, by contrast to his first marriage during which he had eagerly shared his day-to-day concerns with my mother.

Ballistics was to swallow up nearly a quarter of Milne's working life, yet until the archives were made available the only evidence of his outstanding work was a formula which bears his name. In 1870 the Frenchman Jacob De Marre quantified the impact of a shell on armour

---

[2] Letter from Milne to G. Milne, 15 September 1940, EAM, MS.Eng.misc.b.423.
[3] Letter from Milne to G. and K. Milne, 30 October 1940, EAM, MS.Eng.misc.b.423.

plating provided that it struck at precisely 90°. In actuality, of course, shells strike at a variety of angles. Milne modified the formula by incorporating a factor for oblique impact and by making a dimensional correction. The resulting Milne–De Marre equation went into standard use.[4]

At the onset of the war what civilians most feared was attack by poison gas because of its hideously crippling consequences: blistered skin, blindness and even madness. The government issued gasmasks, and anti-gas precautions featured in the course Milne took to qualify as an air-raid warden. (The government wanted one warden for every 500 people.) International law forbade attack by gas, but in the last war this had not deterred the Germans, and the Italians had sprayed the Abyssinians with mustard gas as recently as 1935. If enemy powers chose to violate the Hague Convention, Britain wanted to be able to retaliate. To this end, Milne was directed towards a problem concerning a hollow shell containing poisonous liquid, a vile device that was never deployed. In theory, when the shell struck its target it exploded and released the poisonous, vaporised liquid. But there was an inbuilt snag. In transit the shell failed to keep a steady flight path and veered off-course because the liquid sloshed around inside the cavity and upset the shell's equilibrium. Milne was charged with remedying the instability.

To make the problem manageable, he reduced the number of variables (e.g. by neglecting viscosity) but he knew that his mathematics did not reflect the complexity of the problem. He instigated trials and analysed their results together with data from experiments at Porton Down, Wiltshire, the headquarters of chemical warfare. In a comprehensive report[5] he summarised everything that was known but he was dissatisfied because he could not pinpoint general principles to counter the instability. That he could improve the reliability of any shell, given its calibre, by modifying the shape of the cavity, gave him no consolation despite its obvious practical value. He wanted to delve into the problem more thoroughly, and later, when the 'cold' war posed a threat, he returned to it, but in 1940 he was asked to direct his attention elsewhere. He found switching

---

[4]Longdon, L.W. *Textbook on Ballistics and Gunnery*, War Office 1987, p. 645.
[5]Milne, E.A. 'Report on the Stability of Liquid Filled Shell', External Ballistics Department Report No. 6, O.B. Proc. No. 9492, September 1940, in the author's possession.

problems extremely trying because he disliked treating a problem superficially or having to abandon it half-finished or passing it on to another mathematician so that he could tackle something else. But orders were orders.

A major report like the one Milne had written went to about forty people, many of them within his scientific circle: Lawrence Bragg, Ralph Fowler, Douglas Hartree and Geoffrey Taylor. As the war progressed the number of people in ballistics grew until at its peak the distribution list tripled, with copies of reports going to Washington, Ottawa, Australia, New Zealand, South Africa and India.

While the Allies agreed on the value of co-ordinating research, they had markedly different approaches. The Americans preferred to consider the broad sweep of a problem and this chimed with Milne's outlook. He was at his best when starting from basic principles and reaching general conclusions. He wanted, for example, to look into the fundamental question of whether a shell or a rocket causes more destruction. (Having identified a tangle of factors he began to weigh up their relative merits only to be prevented from going further.) At this critical time Britain was under siege to an extent not known since the Norman Conquest. The civilian death toll was rising and hundreds of buildings had been obliterated. In this state of emergency mathematicians were required to respond to whatever was deemed most urgent and general investigations had to wait. According to changing priorities, Milne had to flit from one assignment to another. For instance, how should a gun be deployed to obtain maximum penetration on a horizontal target? His answer was to decrease the muzzle velocity and increase the gun's elevation. Some problems brought unexpected moments of exhilaration:

> The multiple pom-pom problem has turned out extraordinarily interesting …
> and many of the results … are surprisingly beautiful … They involved what
> to me were original applications of the integral calculus — a field where one
> thought there was nothing left to be discovered![6]

---

[6]Letter from Milne to A.V. Hill, 2 October 1940, AVHL I.

In a fruitful collaboration with John Corner,[7] a mathematician at the Armament Research Department at Fort Halstead, Kent, Milne designed the lightest field gun to meet specified criteria. For example, to penetrate an 8 cm plate at 500 yards (455 m) they concluded that 7 lbs (3.2 kg) was the optimal weight of shot. Sometimes a conclusion was counterintuitive: at long-range and low velocity, a shell with a blunter-nosed head is the most effective.

Mathematicians put forward proposals which were tested in firing trials commissioned by the Ordnance Board. The board issued meticulous instructions to military personnel as to exactly how trials were to be carried out. In this bipartite system stringent regulations, doggedly adhered to by the military, prohibited any modification to instructions in mid-trial, even though results from early rounds in a sequence might clearly indicate worthwhile adjustments. Aside from wasting limited resources, this inflexible procedure infuriated theoreticians like Milne, who were deeply frustrated at having their efforts squandered. If anything were needed to ferment resentment among civilians, it was the absence of military common sense.

Trials that involved several variables might require hundreds of rounds of ammunition. For instance, investigations about the shape of a shell's head depended on five variables and used 500 rounds fired against seventeen types of armour plating. Sometimes the severe shortage of ammunition imposed constraints and led to incomplete trials. As one writer curtly noted, 'There are only six experimental shells available for trials at both high and low striking velocities.'[8]

It was a great relief to get away from the grind of desk-bound mathematics and Milne savoured the change of pace and surroundings when he watched trials at Aberporth in North Wales, at Shoeburyness and on the muddy, exposed Foulness headland. He loved the breezy ferry ride across the Thames between Gravesend and Tilbury. Although visits to military establishments induced a sense of purposeful reality, he made no secret of

---

[7]John Corner (1916–1996). Author of *Theory of the Interior Ballistics of Guns*, New York: Wiley, 1950. Became Chief of the Mathematical Physics Division of the UK Atomic Energy Authority.

[8]Appendix to O.B. Proc. No. 19431, August 1942, p. 3, in the author's possession.

his irritation with petty bureaucracy: 'The number of times I have to sign my name to recover seven shillings and sixpence for subsistence allowance for a spell of over ten hours at Shoeburyness is really ridiculous.'[9]

Milne had an assistant, Norman Hinchliffe,[10] a Yorkshireman with an Oxford First in mathematics, who rated Milne an excellent lecturer. Hinchliffe noticed that when people at Kemnal Manor were at a loss as to how to get a handle on a problem they came to Milne because he was approachable and friendly. Milne invited Hinchliffe home, to tea on Saturday afternoons or sherry on Sunday mornings.

Milne initiated Hinchliffe into the art of research with a problem on the layout of microphones for sound-ranging. The poor weather at Shoeburyness restricted visual trials to the summer months, but if an acoustical method of testing time fuses were adopted this period could be extended. The brief was to ascertain the optimum number and distribution of microphones to collect noise from a shell detonated by a time fuse. The problem was an old one, but Milne took a new approach which incorporated a correction for wind.[11]

Milne's biggest contribution, valid for years to come, concerned the steel plating that encases ships and tanks, whose destruction was of paramount importance in winning the war. The capacity of steel to withstand deformation varies according to its hardness, ductility, tensile strength, plasticity and other properties. Following World War I, when plating was barely half an inch thick (1.3 cm), steel became increasingly stronger, lighter and thicker. The American Sherman tank and the German Panther had 3-inch (7.5 cm) plating while more ponderous tanks had as much as 6 inches (15 cm).

The Allies and Axis powers were desperately competing to manufacture ever tougher armour plating and to devise ever more effective ways of demolishing that of their opponents. Milne was a vital player in this

---

[9]Letter from Milne to G.J. Whitrow, 16 July 1943, GJW, private access granted by GRW, contact Imperial College Archives.

[10]Norman Hinchliffe (b. 1919) continued to work for the Ministry of Supply, later subsumed into the Ministry of Defence.

[11]'The Location of A.A. Shell Bursts by an Acoustical Method', External Ballistics Department Report No. 21, O.B. Proc. No. 14898, November 1941, in the author's possession.

deadly race. Eventually British techniques outstripped the enemy's and our superior anti-tank weapons were decisive in winning the land battles of North Africa and Sicily.

The amount and type of damage that a shell can inflict depends on a myriad of interrelated factors and throws up a host of problems. What calibre, weight and shape of shell are most effective at long range and at short range? What is the best speed for a shell to be travelling at the moment of impact against plating of a given thickness? Why is it that shot striking armour plating sometimes shatters and sometimes penetrates? And an overall question, what is more damaging to a tank, a single large missile or a spray of smaller ones?

Answering these questions deserved more sophistication than Fowler's primitive experiments of firing shot into bundles of telephone directories. Clearly it was necessary to lay down criteria for assessing outcomes of trials, such as distinguishing between shatter and penetration, but before Milne tackled these matters he had to learn some metallurgy.

The National Physical Laboratory (NPL) in Teddington was well advanced in understanding the mechanisms of penetration and the structural changes that occur during deformation. An enterprising young technician Robert Gifkins,[12] who had studied for a degree at evening classes and grew vegetables in the grounds of the NPL, recalled Milne, Geoffrey Taylor and Desmond Bernal coming to see the work of his section. The NPL team had simulated shearing by forcing wooden 'shot' into a rotating Plasticine 'plate' of coloured layers that showed up warp and twist. Grids of lines ruled on the front and back of a target revealed the extent of surface distortion.

For maximum damage, a shell should remain intact while it penetrates and passes through the target, and only as it emerges from the far side should it shatter. For this to occur, the hardness of the shell must gradually decrease from tip to base. Hardness, measured by electrical resistivity, can be altered through heating, and a team at the NPL carefully played gas jets along the length of the shell to adjust its hardness. To lubricate the

---

[12]Robert Gifkins (1918–2006). In 1948 he emigrated to Australia and rose to be Chief Research Scientist at the Commonwealth Scientific and Industrial Research Organisation. He held visiting professorships in Canada and Britain.

moment of impact, the shell's nose was capped with copper, but since copper melts before steel a plastic sheath was added.

The need to create a system of classification for the results of trials gave Milne a rare opportunity to devise a scheme from scratch. He rated each outcome either a 'hole' for a success, or a 'scoop' for a failure, that is, non-penetration. To complicate matters, results were erratic. There was no critical speed above which a hole was a certainty, and below which a scoop was guaranteed. Sometimes a low-speed shot penetrated; at other times, at a much higher speed, it failed to do so. His simple rule to define 'critical velocity' went into routine use, and his statistical approach accommodated peculiar inconsistencies. He invented the 'principle of least contradiction'[13] which states that the 'result is a contradiction if the reverse result is more likely'.[14] Milne was elated with the power of his principle, which he deemed analogous to the principle of least squares that fits the best curve to a scatter of points on a graph. Since his principle dealt with a spread of numerical results, irrespective of context, mathematicians applauded it for its application to other fields such as biology.

He employed statistics to good effect to define the 'ideal' explosion of an anti-aircraft shell. There are three main components: annihilation resulting from penetration, blast of explosion which creates an intense pressure wave, and the harm caused by the casing disintegrating into jagged metal pieces of varying shapes and sizes.

Milne's work on rockets — he stabilised the spin of a rocket — brought him into contact with the eminent statistician Maurice Bartlett.[15] He attributed Milne's 'breakthrough' in the 'ideal' explosion problem to his avoidance of concentrating on fragments of any particular size and his use of a statistical distribution. Bartlett was astonished that Milne rated this analysis among his best in view of his prodigious output in astrophysics and cosmology.[16]

---

[13] 'On the Analysis of Plating Trials', External Ballistics Department Report No. 22, O.B. Proc. No. 15463, Dec. 1941, p. 8, in the author's possession.

[14] Milne, E.A. 'The Determination of "Critical Velocity" in A.P. Trials', Appendix, O.B. Proc. No. 26, 116, WO 195/5684, NA.

[15] Maurice Stevenson Bartlett (1910–2002). Professor at Manchester from 1947 to 1960, at University College London from 1960 to 1967 and at Oxford from 1967 to 1975.

[16] Letter from M.S. Bartlett to the author, 19 September 1990.

Milne imposed order on the phenomenon of 'shatter'. In trying to make sense of its inconsistencies, he evaluated the influence of factors such as striking velocity, thickness of plate and hardness of steel. He instigated trials with a 2-inch (5 cm) gun firing 3 lb (1.4 kg) or 5 lb (2.3 kg) shot, at speeds of 4,250 ft/sec (4,663 km/h) or 3,700 ft/sec (4,060 km/h), into plates from 4 to 10 cm thick. He assigned the outcomes to four categories: shot which rebounded without shattering, shot which shattered on impact making a dent, shot which perforated but failed to shatter, and shot which both perforated and shattered. By displaying the results in a phase diagram he defined criteria for shatter. This led him to suggest suitable tank tactics.[17]

The spring of 1941 brought a fresh wave of terrifying raids, and again Eltham was in the thick of it. While fire-watching one night Milne rushed out to combat flames leaping from a neighbour's garden and carried bucket after bucket of sand to extinguish the fire. As bombs whistled in the distance, he saw the sky fill with menacing orange parachute flares heralding more devastation. Already the war had cost 60,000 civilian lives and another 80,000 had been injured. Conscious that he might be the next fatality, Milne made a new will.

The war was going badly. In May, morale plummeted to a low ebb when HMS *Hood,* the pride of the British navy, was lost with two thousand hands, ostensibly the work of *Bismarck,* sunk two days later by the Fleet Air Arm, though nowadays attributed to *Prinz Eugen.*[18] After the catastrophe Milne was asked to assess the vulnerability of enemy warships. Opinion divided as to the best method of destruction, whether to target the heavily-plated citadel in the centre of a ship with armour-piercing shells or blow up the bridge with high explosives. Much later, in November 1944, Lancaster bombers disposed of the *Bismarck*'s sister battleship *Tirpitz* with 'tall boy' super bombs designed by Barnes Wallis.[19] Demolition of the mighty ship boosted British spirits and reinforced the view that the war was swinging in our favour.

---

[17]Appendices of Proc. No. 20,002 and Proc. No. 24,308, undated, in the author's possession.
[18]Norman, A, HMS *Hood,* New York: Stackpole, 2001.
[19]Sir Barnes Neville Wallis (1887–1979). Engineer. Famous for his 'bouncing bombs' that destroyed Nazi dams.

The hopeful prospect of regaining French and Belgian territory raised a fresh problem which landed on Milne's desk. How best to remove Germany's concrete coastal defences, pillboxes and 'pens' for storing U-boats, which were as much as 20 feet (6 m) thick? A contraption filled with charge called a 'flying dustbin' could blow up thicknesses up to 10 feet (3 m). For greater thicknesses, Milne and Hinchliffe proposed an alternating process, known as 'pick and shovel', which was capable of clearing an aperture big enough for a tank. After a shell broke up the concrete (pick), a high explosive (shovel) cleared away the rubble.

A particularly promising British weapon for destroying tanks was the high explosive squash head. After penetrating armour plating, the head split into a lethal spray of high-speed white-hot ragged pieces that left no survivors. When fired, however, the gun required a special device attached to its muzzle, which immediately afterwards prevented the gun from firing other types of high explosive shell that circumstances might deem to be necessary. A Czech engineer, known as Littlejohn, solved this operational drawback with the armour piercing discarding sabot (APDS), whose distinctive feature was a light metal shoe. As the name suggests, this fell off as soon as the shot emerged from the muzzle, instantly making the gun available for alternative use. Milne improved the accuracy of the APDS by assessing factors like weight, calibre, angle of attack and striking velocity,[20] and he advised on strategy in terms of the most appropriate number of rounds to be fired and the most advantageous distribution of velocity.[21] The APDS proved invaluable and General Montgomery[22] expressed great satisfaction with its performance at Normandy on D-Day.[23]

In 1941, though, before ballistics attained such refinement that the President of the Ordnance Board could assert with authority that we could

---

[20] 'The Performance of Shot Against Plates', External Ballistics Department Report No. 20, O.B. Proc. No. 14981, November 1941, in the author's possession.

[21] 'On the Conduct of Plating Trials', External Ballistics Department Report No. 24, O.B. Proc. No. 16608, March 1942, in the author's possession.

[22] Bernard Law Montgomery, Viscount Montgomery of Alamein (1887–1976). Field Marshal.

[23] Minutes of the 47th Meeting of the Advisory Council held on 27 July 1944, WO 195/6804, NA.

penetrate the enemy's armour plating provided we knew its specification,[24] the outcome of the war hung precariously in the balance. Milne was dejected at the prospect of a long haul with no end in sight. He was overcome with revulsion for the 'orgy of destruction'[25] implicit in ballistics and angry at being instrumental in wrecking bridges and ships, killing soldiers and drowning sailors. To tolerate the inhumanity of his job, he tried to shut out from his mind the brutality of warfare and instead focus on the mathematics. This attitude did not endear him to his 'soldier friends', who regarded mathematics as 'a mere piece of machinery, a sort of spanner for unloosing nuts'.[26]

By sidestepping the 'terrible dreariness of most ballistics calculations'[27] and taking an innovative approach, Milne courted trouble with his 'more narrow-minded'[28] military colleagues, dogmatists who distrusted anything unfamiliar. Anticipating trouble in one of his reports, he faced their antipathy head-on: 'The writer is well aware that some of the above considerations are unconventional … But if nothing new were ever presented … progress would be slow.'[29] As to the dreariness of compiling range tables, one rising mathematician, Derek Taunt,[30] asked to leave Kemnal Manor for more demanding work and became a code breaker at Bletchley Park.

Reginald Jones, no mean raconteur, captured the spirit of an abrasive encounter between Milne and a certain Colonel Kerrison, although the account[31] is skewed. The story goes like this. Clifford Chaffer, Milne's former mathematics teacher at Hymers, asked him to check some calculations done by Kerrison. Milne found them wanting, and added that this was to be expected of a colonel. Incensed by Milne's gratuitous remark, Kerrison informed Milne that he, Kerrison, had beaten Milne to gain the

[24]President's Report of the Ordnance Board 1941–1945. Supplement p. 79, WO 195/362, NA.

[25]Letter from Milne to M.P. Dunham, 5 April 1942, in the author's possession.

[26]Letter from Milne to S. Chandrasekhar, 14 August 1941, EAM, MS.Eng.misc.b.428.

[27]Letter from Milne to W.H. McCrea, 6 December 1941, WHMcC.

[28]Letter from Milne to S. Chandrasekhar, 6 July 1943, EAM, MS.Eng.misc.b.428.

[29]Milne, E.A. 'Expectations, Probabilities and Multiple Hitting', undated, in the author's possession.

[30]Derek Taunt (1917–2004). Cayley Lecturer at Cambridge from 1965 to 1982.

[31]Jones, R.V. *Most Secret War*, London: Hamish Hamilton, 1978, pp. 47–48.

top scholarship at Trinity but had gone to fight in the war. This was sheer fancy. The list[32] of successful candidates in December 1913 does not contain anybody by the name of Kerrison. When taxed with this evidence, Jones pleaded that Milne 'always seemed to have a very mild manner' and that the point of the story was 'the light it threw on Kerrison'.[33] Apparently Kerrison had a reputation for being blustery and boastful, but unless this is made clear the story's message is ambiguous. Nor did Kerrison win an Adams Prize, the punch line in an embroidered version of the story.

Milne's prejudice against the army, which began in World War I, strengthened during World War II. He was not alone in this. The President of the Ordnance Board, Vice-Admiral Francis Pridham,[34] who had been the Captain of HMS *Excellent*, had few good words to say about the army. Charged with shaking up the complacency of service personal at the Ordnance Board, he appointed Milne to be one of two resident scientists; the other was Egon Pearson. As well as recognising Milne's 'splendid work for the Ordnance Board on the piercing of armour by projectiles',[35] Pridham enjoyed chatting to Milne during their lunch hour strolls around the grounds of Kemnal Manor. In their wide-ranging discussions they must surely have talked about Whale Island. Milne's comments on wildflowers and grasses apprised Pridham of Milne's botanical knowledge and when Pridham raised the subject of teaching classics in schools he was left in no doubt as to Milne's enthusiasm for it. Had he not been well grounded in the subject, Milne said, he would be incapable of expressing the abstruse ideas that occur in astrophysics.

Colonel G.O.C. Probert was an exception to Milne's condemnation of bone-headed army officers and he held the colonel in the highest regard, not least for his ingenuity in disposing of a tiresome feature of worn-out guns. Repeated use of a gun deforms its shape and reduces its muzzle velocity. This necessitates re-calculating its firing tables. Why not, suggested the

---

[32]Published in *The Times* and *The Yorkshire Post* on 15 December 1913.

[33]Letter from R.V. Jones to the author, 25 February 1991.

[34]Sir Arthur Francis Pridham (1886–1975), joined the navy at the age of fifteen. Among his many commands was HMS *Hood* during the Spanish Civil War.

[35]Memoirs of Sir Arthur Francis Pridham, p. 198, Prid 2/3, Pridham papers, Churchill College, Cambridge.

imaginative colonel, fashion a new gun so that its dimensions were in accord with those of a worn-out gun? His eminently feasible idea of gradually decreasing the depth of the rifling towards the muzzle became accepted procedure and a gun prepared like this was described as 'probertised'.

By the end of the day, after nine hours at the Ordnance Board, Milne was drained of energy, but domestic chores awaited him. On reaching home he rigged up frames covered in black paper for the blackout, an irritating job because the paper ripped and the drawing pins fell out. He was too tired to muster the drive for his own research, but salvaged his university work from total extinction by taking off the occasional Saturday. Besides meetings at Oxford, he examined doctoral candidates there, and at Cambridge and London, and lectured in Manchester, Liverpool and Edinburgh.[36]

Getting lost in the pitch-black darkness of a chilly, foggy winter's evening was nothing to the joy of a brief immersion in academia. He was disappointed to forgo J.J. Thomson's memorial service in Westminster Abbey but even in wartime he could not go improperly dressed and his funeral clothes were packed away at Northmoor Road. (Tenants occupied all but one room of the house.) As if to compensate, he set his heart on witnessing the installation of Trinity's new master, the social historian George Trevelyan.[37] The 5 am start and six-hour journey were worth every moment in order to be present at the archaic ceremony. The day was rounded off by riding with Newall in his brougham to Madingley Rise where Milne stayed for the night. On other nostalgic visits to Cambridge Milne kept his friendships in repair by calling on the Braggs, the Hartree parents and Lady Rutherford, now a widow.

Milne continued to contribute to *Nature, Philosophy, The Mathematical Gazette* and *The Observatory*. During the tense months of the Blitz he was the target in *The Observatory*[38] of a diatribe from George McVittie, who

---

[36]Milne's James Scott Memorial Lecture, 'The Concepts of Natural Philosophy', delivered on 3 May 1943 is in *Proc. Roy. Soc. Edinburgh* 62, pp. 10–24, republished in Munitz, M.K. (ed.) *Theories of the Universe*, London: Allen & Unwin, 1957, pp. 354–376.
[37]George Macaulay Trevelyan (1876–1962). Master of Trinity College from 1940 to 1951.
[38]*The Observatory* 63, 798 (1940), pp. 273–281.

ridiculed the methodology of kinematic relativity yet could not fault its mathematics. Milne unpicked McVittie's 'misunderstandings'[39] and Walker labelled McVittie's tirade a 'collection of dislikes'.[40] McVittie raised the emotional stakes by accusing Milne of 'thinly veiled sarcasm'.[41] Milne, immersed in his work at Kemnal Manor, felt that McVittie took unfair advantage over him as he had no access to scientific journals for strengthening his case. Eventually Edmund Whittaker soothed Milne with a 'thoroughly decent letter',[42] in which he regretted McVittie harassing him. Yet despite his onslaught — and McVittie would dub Milne the 'stormy petrel of astronomy'[43] — he recognised that Milne exerted 'considerable influence'[44] over him.

Milne was sharply conscious of the adverse comment he attracted and was taken by surprise when the Royal Society awarded him a Gold Medal, a high watermark of British science. On the orders of King George VI to conserve gold, the weighty medal was not struck until after the war (Fig. 20). What Milne most appreciated about the medal was that it made life more bearable by reconnecting him to peacetime colleagues when they wrote to express their congratulations.

Milne missed Oxford's social network. In Eltham he felt an exile, an outsider living a rootless provisional existence devoid of any sense of community. He put aside Sunday afternoons for the family, either taking a walk on Chislehurst Common, Avery Hill Park, Petts Wood or Shooters Hill, or going on an expedition by bus to Knole Park or to Greenwich for tea with Harold Newton,[45] an authority on solar flares and magnetic storms. Julia Mann,[46] the Principal of St Hilda's College, Oxford — where

---

[39] 'Kinematic Relativity: A Discussion' in *The Observatory* 64, 800 (1941), p.11.

[40] 'Kinematic Relativity: A Discussion' in *The Observatory* 64, 800 (1941), p.17.

[41] 'Kinematic Relativity: A Discussion' in *The Observatory* 64, 800 (1941), p. 25.

[42] Letter from Milne to A.G. Walker, 30 November 1942, AGW.

[43] 'English Astronomer E.A. Milne Dies Suddenly' in *Sky and Telescope* (November 1950), p. 13.

[44] G.C. McVittie Autobiographical Sketch, Niels Bohr Library and Archives, American Institute of Physics, College Park, MD, USA.

[45] Harold William Newton (1893–1956), worked at the Royal Observatory from 1910 to 1955.

[46] Julia de Lacy Mann (1891–1984). Economic historian. Principal of St Hilda's College, Oxford, from 1928 to 1955.

Milne served on the council — treated them to outings in her car when she visited her mother in Bromley until petrol rationing put a stop to that. Margot's sister Jean and her husband had the Milnes for the odd weekend to their large comfortable house in the New Forest, where Milne revelled in the luxury of browsing among their books. He enjoyed reading a book on the life of Socrates.

Work and the struggle to survive dominated Milne's life. Food was dreary with shortages of imported items such as rice, chocolate and coffee. Bananas, lemons and oranges were non-existent. Manufactured goods like soap, camera film, string and razor blades were hard to come by because factories had been turned over to the production of munitions. There was no possibility of 'shopping around' because everybody had to register with a grocer for weekly rations of butter, bacon and cheddar cheese (called 'mousetrap'), and with a butcher who would sell 'off-the-ration' extras, like offal, only to his registered customers. Dried egg was the norm — Alan called Humpty Dumpty a 'shell egg'. The six eggs sent by Margot's father in Dornoch, miraculously arriving intact, were a great treat. Food parcels from the Chandrasekhars and Walter Adams, Edwin Hubble and staff at Mount Wilson Observatory brightened the monotonous diet.

Once when Beatrice was unwell she dispatched Milne to the shops with a perambulator. Nothing better encapsulates the change in the social order than the image of Milne pushing a pram filled with shopping, unthinkable for a professional man in pre-war days. 'Shopping is a great bother … the shops are terribly crowded',[47] he wrote after standing in line to watch the sole piece of cheese get smaller and smaller, and finally be allowed just 4 ounces (114 grams). Three months later, 'Meat is getting very scarce here, and we depend largely on carrots and potatoes. Beatrice made a Lord Woolton[48] pie yesterday',[49] a concoction of root vegetables promoted by the government and named after the popular Minister of Food, Uncle Fred.

---

[47]Letter from Milne to the author, 1 February 1941.

[48]Frederick James Marquis, Earl of Woolton (1883–1964). Social reformer, businessman and politician. Minister of Food from 1940 to 1943. Minister of Reconstruction from 1943 to 1945.

[49]Letter from Milne to the author, 20 May 1941.

Milne was appalled that his annual leave was cut to a single week. For the precious holiday the Milnes went to the pretty village of Ampleforth, in the depths of Yorkshire, which he knew well from visits with the Fiddeses. Amazingly, rationing had hardly impinged at all in the area, and food was plentiful. Milne showed Beatrice Helmsley Castle and the Norman ruins of Rievaulx Abbey. He loved the blissful tranquillity of wandering on the moors among the heather and bracken, away from the cares of his never-ending work.

In the autumn of 1941 Milne eagerly welcomed Theodore Dunham, who was in England on behalf of the American government. Vannevar Bush,[50] President Roosevelt's adviser on military technology, had recruited Dunham to improve optical devices. (Bush was President of the Carnegie Institution, which included Mount Wilson, and knew of Dunham's expertise in spectroscopy.) Dunham was in Wales flying with the RAF to test gun sights, binoculars and cameras when news broke of the Japanese attack on the American naval base, Pearl Harbor. The officers in the Mess burst into spontaneous clapping and Dunham's first reaction was one of shock until he realised that they were applauding the raid because it would bring America into the war and strengthen the Allies. The European war was now a global war.

During Dunham's visit, the Milnes moved to Mottingham. The house at 22 West Park was more agreeable than their Eltham flat. It was near the shops, the garden was big enough for Beatrice to grow vegetables and it had a spare bedroom. One Christmas the Milnes were glad to have Bill McCrea with them when he could not get home to Belfast. He thought Beatrice was highly strung.

By 1942 the war was at a grim stage. Besides defeats in Greece and Africa, the Allies had suffered devastating shipping losses in the Mediterranean, the Atlantic and the Pacific. The national mood was sombre and the fabric of life deteriorated: conditions were more wretched, clothes more threadbare, the dismal diet more limited. Even apples were

---

[50]Vannevar Bush (1890–1974). Engineer. Administrator. He supervised 6,000 researchers in 300 laboratories. At the Massachusetts Institute of Technology he built a differential analyser, the precursor of D.R. Hartree's machine in Manchester.

scarce. Milne had recently reported that 'milk is getting short here now because so much of the land is ploughed for corn'.[51]

To add to the widespread despondency, Milne grieved for his brother Geoffrey,[52] who died in acute pain from an infection of the pancreas. His sudden death hit Milne hard. From childhood they were soulmates and together they had shouldered the family burdens. Milne was concerned for his widow and two sons stranded in East Africa and spent sad evenings attending to their affairs.

This war, unlike the last, held no sparkle for Milne. Moreover he was plagued by poor health. His ailments encompassed gastroenteritis, 'flu', boils, an abscess lanced under gas and most seriously chest pains from incipient heart disease. His teeth were painful and after they were extracted the incompetent dentist fitted his dentures badly and they rattled about causing ulcers.

While looking in the mirror one day, my father noticed that his face was old beyond his years. He was aware, too, that he spoke less fluently. These were the delayed sequelae of *encephalitis lethargica*, progressive symptoms of Parkinson's disease that included a tendency to drop off to sleep at odd moments. Hinchliffe monitored his lapses with little tests of alertness. If they were proofreading with Milne checking and Hinchliffe reading aloud, he deliberately introduced artificial errors to see whether Milne was awake.

As the war rumbled on, Milne felt jaded, over-worked and undervalued. Above all, he was exasperated by the government's cavalier attitude to civilian mathematicians, contemptuously treating them as robots pumping out solutions, rather than human beings wrestling with difficult problems that required initiative, inspiration and perseverance. There was no understanding that mathematicians need fallow periods in order to attain crescendos. J.B.S. Haldane reckoned that no one who engages in mathematics for more than five hours a day can remain sane.[53] The government was short of mathematicians, yet instead of nurturing them, it

---

[51]Letter from Milne to E. Milne, 30 November 1941, in the author's possession.
[52]An obituary of Geoffrey Milne is in *Nature* 149 (14 February 1942), p. 149.
[53]Clarke, R. *J.B.S. Haldane*, London: Hodder & Stoughton, 1968, p. 29.

callously imposed gruelling hours that eroded morale, stultified motivation and reduced productivity.

Milne was not given to outbursts of fury, but he reached the end of his tether at the government's 'slave driving'[54] demand that mathematicians extend their five-and-a-half-day week to six. The diktat impelled him to relay the grievances of mathematicians to A.V. Hill, now an independent Member of Parliament, aware that the letter might be construed as seditious:

> It is impossible to do first-rate research or investigation of a mathematical character at the rate of 51 weeks out of 52, week in week out, mornings and afternoons alike. After a spell of exacting white-heat research, the mind employed in mathematics goes slower and slower, and sooner or later initiative drops, progress stops and the very thought of engaging in mathematics becomes nauseating. Originality flags. ... Mathematical work is ... different from the work involved in making decisions, or attending trials ... all of which afford relief from the severity of strained application. ...
>
> Cabinet Ministers pay lip service to research, but they are blind to the circumstances which destroy initiative. ... A large number of junior workers ... have no opportunities of such relaxations as attending trials or meetings or travelling about, and it is their staleness which makes the State the loser.[55]

Hill sent copies of the letter to R.A. Butler[56] and to Sir Henry Dale,[57] President of the Royal Society, adding a note that scientists' salaries were under review. The pay for junior mathematicians was disgracefully low. Certainly Hinchliffe resented his pittance of £185 per annum. Oxford University topped up Milne's remuneration from the Ministry of Supply to his pre-war salary, but life in London was more expensive than Oxford.

---

[54]Letter from Milne to T. Dunham, 24 June 1941, in the author's possession.
[55]Letter from Milne to A.V. Hill, 25 March 1942, Dale papers, Royal Society, London.
[56]Richard Austen Butler, Baron Butler of Saffron Walden (1902–1982). Politician.
[57]Sir Henry Hallett Dale (1875–1968). Physiologist. President of the Royal Society from 1940 to 1945, of the Royal Society of Medicine from 1948 to 1950, and of the British Council from 1950 to 1955.

When the ministry did award him a rise, he grumbled that he reaped no benefit since it merely diminished the university's contribution.

Undoubtedly Milne's close proximity to military personnel exacerbated his discontent. Officers at the Ordnance Board were exposed to no more danger from bombing raids than he was. Their leave allocation was more generous than his and they did not have to subsist on civilian rations. (Jean's husband, Edwin, who had piloted his own plane in peacetime and was an RAF instructor, always started his day with an egg.)

That Milne was acutely aware of his civilian status at the Ordnance Board is evident from the following episode. The board wished to send a greetings telegram to the American Embassy for Independence Day and the secretary called for suggestions. 'The Naval and Military Members of the Board were dumb ... eventually I, the only civilian present, ventured on a form of words, which was accepted with minor changes.'[58] Milne earned a reputation for versatility and one senior officer claimed extravagantly that Milne could crack an intractable ballistics problem, explain a forgotten theorem of Euclid, recite an ode from Horace or a speech from Shakespeare, and tell a bawdy joke, all in the same lunch hour.[59]

Despite his immersion in ballistics, Milne had not abandoned his quest to derive the laws of nature and was determined to set down ideas that were bubbling in his mind before his ability 'froze up altogether'.[60] To do this he rose early in the morning and put in a couple of hours before going to Kemnal Manor. The result was five papers[61] on electrodynamics that he wrote during the autumn of 1942 and which he credited with keeping him 'mentally alive'.[62]

Later that year he joined the Advisory Council of Scientific Research and Technical Development, the highest body of the Ministry of Supply, whose members were top scientists such as Desmond Bernal, John Cockcroft, Charles G. Darwin, Robert Robinson, Richard Southwell and

---

[58]Letter from Milne to M. Dunham, 5 July 1943, in the author's possession.

[59]Letter from Brigadier P.S. Gostling to W.H. McCrea, 13 November 1950, EAM, MS.Eng.misc.b. 423.

[60]Letter from Milne to T. Dunham, 17 October 1942, in the author's possession.

[61] 'Rational Electrodynamics I to V' in *Phil. Mag.* 34, 7 (1943), pp. 73–82, 82–101, 197–211, 235–245, 246–258.

[62]Letter from Milne to O. Struve, 29 April 1944, Struve papers, Yerkes Observatory.

Henry Tizard. Members were chosen as individuals on personal merit, irrespective of their employment, and Milne was admonished for asking for Advisory Council papers to be sent to Kemnal Manor rather than his home.

The council took vital decisions about assigning limited resources to competing priorities based on information filtered up through a labyrinth of committees and sub-committees. Milne already sat on some of these and opposed further proliferation. He blocked the creation of a new one on his own speciality, armour plating. By an extraordinary anomaly, which stretches credulity, the council had no link with the Ordnance Board, because the board, anxious to preserve its historic independence, refused to be swallowed up in the government's bureaucratic machine. Consequently the council was not up-to-date with the board's latest activities. Milne became the informal conduit of communication and, for instance, reassured the council that, yes, the effect of wind on the spin of a rocket had been quantified.

Just as his brain began to curdle, barely capable of coming up with fresh mathematical ideas, he increased his administrative responsibilities by becoming chairman of the Ballistics Committee. His first move was to streamline its remit by integrating its three strands: internal ballistics, which deals with the inside of a gun; external ballistics, which deals with the missile in transit; and terminal ballistics, which deals with the missile's impact on the target. Milne was fortunate that his secretary was Frank Smithies,[63] a former pupil of G.H. Hardy who had explored Milne's integral equations and who would become a Cambridge pillar of pure mathematics. Being secretary to several committees, Smithies was an invaluable go-between and was accustomed to working with chairmen of varying temperaments. He had an easy relationship with Milne, and, apart from occasional fussiness, noticed that Milne, unlike some chairmen, did not bully members or overly push his own opinions.

By the end of 1943 the collapse of Rommel's army in North Africa, the landings in Sicily and the surrender of the Italians, decisively tipped the tide of the war towards the Allies. To add to this surge of optimism, Milne rejoiced in his new daughter Edith, born in February. She was a

---

[63] Frank Smithies (1912–2002). Fellow of St John's College, Cambridge, for sixty years.

revelation to him for the care an infant needs and the labour it entails. His first crop of children had been at one remove in the nursery and Edith gave him hands-on experience, as he now helped with chores such as the washing, which was done by hand and without a mangle. He doted on Edith, took inordinate pride in her development and was pleased to see that she softened Beatrice's severity.

Shortly after her birth, Milne was almost as proud of earning an Oxford doctorate in science by submitting material to examiners in the normal way and he hoped the legitimate degree would make him a respectable mathematician. (He discounted his wartime BA from Cambridge, his courtesy MSc from Manchester and MAs from Oxford and Cambridge.) At Oxford a doctor ranks higher than a professor and carries the privilege of participating in Encaenia and other ceremonies at the Sheldonian Theatre. Bedecked in his scarlet gown, he could now process in stately splendour alongside other senior members of the university.

A few months later he became President of the Royal Astronomical Society, which he found more exacting than the presidency of the London Mathematical Society. Only two other people had held both positions, Sydney Chapman and James Glaisher.[64] All three were Trinity mathematicians and all three had connections with research on the Earth's atmosphere, Glaisher through his father, the balloonist James Glaisher.[65] During Milne's presidency the RAS saw fit to elect the first woman to its council. Madge Adam,[66] petite, wiry and self-deprecating, was the first woman to achieve an Oxford First in physics and throughout the war she ran the OUO while Plaskett was occupied with navigation instruments. As a contribution to the war effort, the RAS joined with the British Astronomical Association to put on a series of 'popular' lectures for members of the armed services stationed in London. On a raw December night Milne and

---

[64] James Whitbread Lee Glaisher (1848–1928). Lecturer and Fellow of Trinity College from 1871 to 1928.

[65] James Glaisher (1809–1903). Astronomer and meteorologist. Secretary of the Royal Meteorological Society from 1850 to 1872.

[66] Madge Gertrude Adam (1912–2001). For twenty years she directed physics teaching in all of Oxford's women's colleges. Twice she was Vice-President of the RAS.

Jeans stumbled through the blackout to Burlington House where Jeans gave the opening lecture on 'Photographic Astronomy'.

In the winter of 1943/1944, bombing on London intensified. 'We have been lucky ourselves so far, but any night we may be the destination of an incendiary or other bomb.'[67] Taking a crawling baby on holiday was out of the question, and Milne spent his annual leave in the solace of the RAS library, where, among other things, he prepared his presidential address.[68] Afraid that a raid might destroy his manuscript, he took the precaution of sending a copy to a printer outside London.

On 15 June 1944 began the scourge of the terrifying V-1s. The V stands for Vergeltungswaffe, reprisal weapon, used in retaliation for the D-Day landings in Normandy the previous week. The aerial torpedo, with a flaming tail and a wingspan of 6 yards (5.5 m), was expressly designed to kill civilians. At a pre-set time the fuel supply cut out, the engine stopped and the 'doodlebug' crashed to the ground detonating a ton (1,016 kg) of explosives. Just before the crash the ominous silence was a frightening, heart-stopping moment as it meant certain death for those below. The terror of the doodlebugs was immeasurably worse than that of a bombing raid. Sirens gave warning of a raid, which lasted for a finite duration and ended with the wail of the 'All Clear'. By contrast, the V-1s appeared at 'any hour of the day or night, fine or cloud. One hears their characteristic buzz and hopes it will last till they get beyond over-head. Then one hears the buzz sharply cut off, and a few seconds later a sickening bang.'[69]

Through an error by the Germans that British Intelligence took care not to reveal, many V-1s fell short of their target of central London and landed in south London. Eventually radar-controlled proximity fuses activated shells that intercepted doodlebugs, but even so, some 6,000 civilians were killed and 18,000 injured. Mottingham was razed to rubble and four men died at the nearby War Memorial. The Milnes were in constant jeopardy and twice Beatrice had a narrow miss. After blast damage blew out their

---

[67]Letter from Milne to M.P. Dunham, 26 March 1944, in the author's possession.

[68]'On the Nature of Universal Gravitation' in *Mon. Not. of the Roy. Ast. Soc.* 104 (1944), pp. 120–136.

[69]Letter from Milne to M.P. Dunham, 17 July 1944, in the author's possession.

windows, the Milnes were advised to leave. Alan was sent back to the Children's Home in Buckinghamshire, where he had stayed for Edith's birth. E.H. Rayner's sister-in-law agreed to take the Milnes into her house in Teddington for three weeks as paying guests and on 3 July they left Mottingham with a few necessities, expecting to return before long. They never did because on 20 July a V-1 wrecked their house.

Despite the crisis, Milne did not neglect his work. At a Scientific Advisory Council meeting, he was apprised of General Montgomery's comment that 'we were now in advance of the enemy in the design of anti-tank weapons',[70] surely a heartening recompense for his endeavours.

My father telegraphed the Dunhams to tell them of the disaster, but I was kept in ignorance and even if I had been told I doubt whether I would have grasped how dire was their situation. Their ruined house was a dreadfully upsetting sight, a mess of soot and rubble, the roof off, water leaking, chunks of ceiling, glass and wood all over the place, 'books and papers thrown out of bookcases'.[71] Heroically Beatrice camped in the cellar to rescue what she could from the debris, and after filling in many forms my father was granted a petrol permit to enable him to hire a man with a van to take their salvaged belongings to 19 Northmoor Road for storage. A trifle begrudgingly their landlady allowed the Milnes to stay on for just one extra week. Homeless, with a wife and toddler, my father was at his wit's end, frantically trying to find somewhere, anywhere, to live, but he found nothing.

---

[70]Minutes of the 47th Meeting of the Advisory Council held on 27 July 1944, WO 195/6804, NA.

[71]Letter from Milne to M.P. Dunham, 23 July 1944, in the author's possession.

Chapter 16

An Invitation

> *Oh Thou, who didst with vodka and with gin*
> *Beset the road they were to wander in,*
> *   Ask not that Bernal, Darwin, Blackett, Mott*
> *Shall spill the atomic beans in alcoholic sin.*
>
> *Requirements of the conflict with Japan*
> *And no intent to scramble men with men ...*
> *   And others, Norrish, Rideal, Milne, Dirac ...*
> *For them, alas! No glory of the also ran.*[1]

Shaken and perplexed, and probably demoralised, the Milnes had little choice but to take temporary refuge with Edith Milne. For Milne to continue with the Ordnance Board he had to be within commuting distance of its headquarters, now at 34 De Vere Gardens, Kensington, and he decided to take lodgings in Harpenden, Hertfordshire, near his brother Philip. Work at the Ordnance Board was much diminished because it was most unlikely that there would be enough time before the war ended to devise and manufacture any new artefacts resulting from research.[2] Some of Milne's civilian colleagues were back at their universities and he expected to be released soon. After his long absence from Oxford, he was champing to get back into harness. To expedite his release he enlisted the aid of Douglas Veale, the university registrar, whose early career in the higher echelons of the civil service equipped him to deal with officialdom. In the autumn the Ordnance Board released Milne from full-time employment and he reverted to his former arrangement of working part-time.

---

[1] Verses from A.V. Hill's unpublished poem 'Backroom Boys'. Copy in the author's possession.
[2] Letter from Sir Henry Dale to C. Attlee, 4 February 1944, CAB 127/213, NA.

He was more than content to return to Oxford to be among its familiar buildings and have the company of colleagues. He lived in Wadham in a set of rooms that had windows onto the road and onto the front quad. By making the college his home he fulfilled the maxim that being a don is not so much a job as a way of life. Frayed and war-weary, he was thankful to be in the folds of the university and able to dispose of his time as he chose, not at the behest of military regimentation.

He restarted his colloquia and resumed lecturing. After World War I he had thrown himself into reviving Cambridge clubs and societies, and now he resuscitated Oxford's University Mathematical and Physical Society, moribund since 1939. He drew up a list of potential members and organised speakers. One of these was David Evans,[3] a future historian of astronomy, who, through peculiar circumstances, gained a Cambridge PhD while working at the Oxford University Observatory. It came about like this. After one postgraduate year at the Cambridge Observatory, his supervisor Richard Woolley[4] moved to Australia. To continue in his particular line of study, Evans took a job as a research assistant at the Oxford Observatory, only to discover to his dismay that Oxford University would not allow him credit for his Cambridge year. Conversely, the more enlightened Cambridge authorities were willing to accept work done at Oxford. This had the bizarre effect that Milne, who was one of his examiners, had to travel to Cambridge, as did Evans, for his thesis interview.

Meanwhile Beatrice endured a ghastly time. With great hospitality the Plasketts took her and little Edith into their Oxford household while she searched for accommodation to tide them over until the tenants at 19 Northmoor Road vacated the house. Her fierce independence, however, created friction and she was obliged to leave. She took Edith to a primitive cottage in Brightwell Baldwin, twelve miles outside Oxford, where Alan joined them. They had a wretched existence. In the unhygienic conditions, without running water, gas or electricity, she was taxed to the limit looking after the children. Their sole source of heating was paraffin stoves and she had to drag drums of paraffin from the village. Obtaining water was

---

[3] David Stanley Evans (1916–2004), astronomer and author, held positions at the Radcliffe and Royal Observatories in South Africa and at the McDonald Observatory, Texas.
[4] Richard van der Riet Woolley (1906–1987). Astronomer Royal from 1955 to 1966.

equally arduous. Having pumped it up from a well, she lugged it to the cottage, but she needed more, and resorted to using rainwater that she had to filter. At weekends Milne came out by bus and tried to help but the heavy lifting made his heart race. Often he found one or all of his family ill with gastric upsets.

In December the Milnes regained possession of 19 Northmoor Road, and almost immediately Beatrice went to pieces. Throughout the terrors of the Blitz and doodlebugs, the privations of the war and the hardships of the cottage, she was resourceful, resilient and courageous, but once these challenges were gone her composure deserted her. Overwrought and unnaturally talkative, she did housework far into the night. Apparently she had suffered previous episodes of instability and her family, who were concerned about her, wondered whether Milne might be to blame for triggering another episode. Through diplomatic connections they knew the American Ambassador, Gil Winant,[5] and prevailed on him to investigate. (Beatrice's brother-in-law had been an honorary Vice-Consul in Ecuador and Peru, and one of her cousins was a Vice-Consul in Edinburgh.) Apparently Winant found nothing amiss.

When Beatrice turned violent she had to be admitted to hospital. In this latest emergency, Milne was grateful to the Red Cross for looking after the children for two days before driving them to the Children's Home, now at Worthing on the Sussex coast. A month later the doctors allowed Beatrice to come home but she was unpredictable and one morning she disappeared without a word. Alarmed by her absence, Milne notified the police, who tracked her down to the Eltham nursing home where she gave birth to Edith. Sydney Chapman's wife, Katharine, collected her and returned her to the Oxford hospital. Alone in the empty house Milne felt utterly miserable. Wracked with worry, all hopes of normal living crumbled; he needed the utmost tenacity to prepare his presidential address for the RAS.

He put on a good front for the Dutch-American astronomer Gerard Kuiper,[6] a friend of Theodore Dunham, who came to see how he was

---

[5]John Gilbert Winant (1889–1947). American ambassador from 1941 to 1946.

[6]Gerard Peter Kuiper (1905–1973). Director of the Yerkes Observatory from 1947 to 1949 and 1957 to 1960. He gives his name to the Kuiper belt, a ring of icy objects at the extremity of the Solar System.

getting on. They met at the Mathematical Institute and Milne took him home for tea. Kuiper thought that although Milne was 'somewhat fragile and aged',[7] he was cheered by his visit and by news of my sister and me.

Milne buried himself in work. His vector textbook with Chapman had lain dormant for the duration of the war. After spending a weekend working on it together, Milne concluded he would have to do it on his own. This did not mar their relationship, and when Oxford's Sedleian Chair became available, Milne persuaded Chapman to accept it, a move Chapman came to regret. Lord Cherwell seemed to bear him a grudge,[8] and Chapman deplored the way the university treated its professors. At Imperial College he was accustomed to ample secretarial assistance, whereas at Oxford he had none and Katharine learned to type despite having arthritis in her wrists.

Milne was glad to oblige his old friend from Cambridge days, Martin Johnson,[9] with a foreword for his book *Time, Knowledge and the Nebulae.*[10] Slightly built, modest and briefly Milne's pupil, Johnson placed science in the sweep of human experience and, like Milne, wished to unify it with religion. They shared a belief in the importance of time in the world of physics. Johnson admired Milne's work[11] and the book is a semi-popular exposition of kinematic relativity and the two time scales. Inevitably this made Johnson vulnerable to Dingle's barbs. In vain Johnson asked Dingle to stop attacking Milne. Dingle had thought highly of Johnson's previous book, *Art and Scientific Thought,*[12] which had an introduction by Walter de la Mare[13] who enjoyed discussing with Johnson the interplay between science and art, and consulted him before publishing his poems.[14]

---

[7] Letter from G.P. Kuiper to T. Dunham, 12 March 1945, in the author's possession.

[8] Letter from H.H. Plaskett to H. Spencer Jones, 26 March 1941, papers of Harold Spencer Jones MS.RGO. 9/553, Cambridge University Library.

[9] Martin Christopher Johnson (1896–1983). Lecturer and reader at Birmingham University. In his 1952 Eddington Memorial Lecture he gave a critical assessment of Milne's time scales.

[10] Johnson, M.C. *Time, Knowledge and the Nebulae*, London: Faber and Faber Ltd, 1945.

[11] Letter from M.C. Johnson to Milne, 20 June 1945, in the author's possession.

[12] Johnson, M.C. *Art and Scientific Thought*, London: Faber and Faber Ltd, 1944.

[13] Walter de la Mare (1873–1956). Poet, novelist and short story writer.

[14] Noted in Whistler, T. *Imagination of the Heart*, London: Duckworth, 1993, p. 377.

In early 1945 J.B.S. Haldane re-awakened interest in the time scales. He showed that by extrapolating mathematically backwards in time, the Solar System could have been 'born' from a collision between the Sun and one or more massive primordial photons of enormously high energy.[15] The nub of his argument, dubbed by Desmond Bernal the 'Milne–Haldane' hypothesis,[16] was that when the Universe was created (at t = 0), there was an optical beginning as well as a dynamical one. If similar collisions had created planetary systems elsewhere, these, too, might support life. Haldane's train of thought prompted Milne to suggest that high frequency cosmic rays might be relics of primitive radiation from long ago.[17]

As usual, like a dog gnawing a favourite bone, Dingle scoffed[18] at anything connected with kinematic relativity and warned Haldane not to waste 'even five minutes'[19] of his time on Milne's proposal. Haldane disagreed. He reckoned it was 'worth a few weeks',[20] and in an American article[21] ruminated on biological conditions thousands of millions of years ago. By calculating the amount of energy a dinosaur needs, he showed that the Universe was more propitious for supporting life than had been previously supposed.

During this lonely period for Milne, the Astronomical Society of the Pacific awarded him its Bruce Medal. The recipient is chosen by the directors of six major observatories, three in America (Harvard, Lick and Yerkes) and three in Europe (Cordoba, Paris and Greenwich), and consequently the medal carries international status, which accurately reflected Milne's standing. To allow time for the society to ship the medal from San

[15]Haldane, J.B.S. 'A Quantum Theory of the Origin of the Solar System' in *Nature* 155 (3 February 1945), pp. 133–135.

[16]Bernal, J.D. 'The Birth of the Planets' in *The Daily Worker* (10 April 1945).

[17]'The Ageing of Light' in *Nature* 155 (24 February 1945), p. 234.

[18]'Kinematic Relativity and the Nebular Red-Shift' in *Nature* 155 (28 April 1945), pp. 511–512.

[19]Letter from H. Dingle to J.B.S. Haldane, 19 January 1945, Dingle papers, Imperial College.

[20]Letter from J.B.S. Haldane to H. Dingle, 24 January 1945, Dingle papers, Imperial College.

[21]Haldane, J.B.S. 'A New Theory of the Past' in *American Scientist* 33 (1945), pp. 129–188.

Francisco to England, it asked him to keep the news to himself until formally announced in May.

In May, Beatrice's doctors reported that she was making progress, albeit slow, and Milne felt he could safely accept an enticing invitation from the Soviet Academy of Sciences, all expenses paid, as part of a prestigious British delegation. After the strictures of the war Milne was thrilled at the prospect of getting abroad, glimpsing Soviet science and fraternising with the other delegates.

As soon as the war in Europe had ended on 8 May 1945, the Soviets wasted no time in setting up a huge international conference in Moscow and Leningrad on the pretext of celebrating the 220th Jubilee of the Academy of Sciences. Delegates from eighteen countries would assemble with hundreds of Soviet scientists for a magnificent two-week extravaganza of unimaginable grandeur.

The Academy of Sciences was the largest scientific organisation in the world and controlled a huge network of institutes. Peter the Great planned the academy when Russia had no universities, and, wishing to emulate the Western study of science, he visited seats of learning in Northern Europe. He met Edmund Halley and Isaac Newton, President of the Royal Society, and while the academy differs in its remit from the Royal Society, they maintained cordial relations. So it was fitting that the Royal Society, on behalf of the British government, handled responses to an event which betokened future amicable relations.

On 13 June Milne packed his bag, locked the house and went to London, anticipating an enjoyable break from his troubles. He was looking forward to a considerable adventure. Early the next morning he was to fly to Leningrad in the plane sent by the Soviets to fetch the British contingent, led by Sir Robert Robinson, President-Elect of the Royal Society. Milne knew most, if not all the delegates: predominantly physicists and chemists, a few biologists, two scholars from the British Museum and the socialist historian Richard Tawney.[22] At the farewell reception in the Royal Society's rooms there was a palpable air of excitement. The trip had assumed an official mantle and among the guests were senior diplomats,

---

[22]Richard Henry Tawney (1880–1962). Historian and social reformer. Lecturer, reader and professor at the London School of Economics.

Robert Barrington Ward,[23] and Lord Cherwell. Also present was Sir John Anderson,[24] the distinguished Chancellor of the Exchequer, who would initiate the drama that was about to unfold.

Sir John Anderson, inventor of the low-cost air-raid shelter that bears his name, was the politician whom Churchill trusted above anyone else. Should he and Anthony Eden fail to return from overseas missions, Churchill stipulated that Anderson was to take charge of the country. Anderson was an accomplished chemist — he had investigated the radio-active properties of uranium — and was the indispensable link between government and science. He was the only member of the cabinet to understand nuclear fission and he obtained the raw materials needed to manufacture the atomic bomb. During its secret development he nursed diplomatic relations with the Americans.

To his astonishment Milne was ushered into a side room along with seven other delegates: the physicists Desmond Bernal, Patrick Blackett, Charles G. Darwin, Paul Dirac and Nevill Mott,[25] and the chemists Eric Rideal[26] and his former pupil Ronald Norrish.[27] There they found Sir John Anderson waiting for them. He told them that their exit permits, without which they could not leave the country, were cancelled, and they were forbidden to go to the USSR. The group were utterly dumbfounded because their exit permits had only just been issued. Milne's was stamped the 9th of June. Peremptory cancellation on the eve of departure seemed inept beyond belief.

John Cockcroft, at work on the atomic bomb in Canada, was put under a similar embargo. His invitation from the Soviets was forwarded to him from England with a note attached, 'Attendance not permitted by British authorities'.[28]

---

[23] Robert McGowan Barrington Ward (1891–1948). Editor of *The Times* from 1941 to 1948.
[24] John Anderson, first Viscount Waverley (1882–1958). Administrator and statesman.
[25] Sir Nevill Francis Mott (1905–1996). Appointed professor at Bristol in 1948 and at Cambridge in 1954. Master of Gonville and Caius College from 1959 to 1966. Nobel Laureate 1977.
[26] Sir Eric Keightley Rideal (1890–1974). Colloid chemist. Appointed professor at Cambridge in 1930 and at the Royal Institution in 1946.
[27] Ronald George Wreyford Norrish (1897–1978). Physical chemist. He spent his life in Cambridge, appointed professor in 1937.
[28] The papers of Sir John Cockcroft, CKFT 11/6, Churchill Archives Centre, Cambridge.

As the extraordinary news filtered round the room, the reception degenerated into a debacle. The atmosphere became highly charged with emotions ranging from disappointment and dismay to resentment and anger. The remaining delegates were of such a disturbed frame of mind that they considered refusing to go on the trip, but they did depart as planned. On arrival, Peter Kapitsa greeted them by enquiring why Bernal, a frequent visitor to the USSR, was not among them. A Soviet official pertinently remarked, 'So Churchill wouldn't let the physicists come here.'[29] Milne returned home feeling decidedly dejected. On the doormat he found a personal letter of apology from Lord Woolton, who intimated that Milne's war work was too important to be put aside. Since Milne had just completed a full academic year at Oxford, this far-fetched explanation did not carry the ring of truth.

It emerged that these eight scientists were detained on the orders of the Prime Minister for reasons which he could not divulge.[30] Had the exit permits never been granted in the first place there might have been less public outcry. As it was, the media made capital out of the fiasco, denigrating a government of waning popularity for its clumsiness and demanding to know precisely why the scientists deserved this restriction.

A.V. Hill, Foreign Secretary of the Royal Society and an ardent supporter of international science, was outraged. He vented his exasperation in an indignant letter to *Nature*,[31] castigating the government for its 'studied discourtesy' and 'offensive' treatment of scientists, phrases picked up by the press. He persuaded the well-known back-bencher Tom Driberg[32] to probe the circumstances of the 'banned eight' by asking a private notice question to the Prime Minister in the House of Commons. Churchill prevaricated:

His Majesty's Government are happy to know that it has been possible for twenty-one of the scientists invited from this country to undertake the visit. ...

---

[29] Ibid.

[30] Gowing, M. *Britain and Atomic* Energy, Oxford: Macmillan, 1964, p. 359.

[31] *Nature* 155 (23 June 1945), p. 753.

[32] Thomas Edward Neil Driberg, Baron Bradwell (1905–1976). Journalist and politician. Independent Member of Parliament from 1942 to 1945 and Labour Member from 1945 to 1974.

in the case of eight other scientists who had accepted, His Majesty's Government did not feel able to authorise the grant of facilities for the journey. His Majesty's Government on consideration found that it was impossible to spare from the United Kingdom at this stage of the war against Japan ... so many eminent scientists whose services they may wish to employ ...

Driberg pressed:

> Is the Prime Minister aware that the scientists who have been refused permission to leave include such world famous men as Professor Milne, Professor Norrish, Sir Charles Darwin and Mr Bernal, and that these scientists have been told by the Ministry of Information ... that it was only the last minute interference of the military security authorities which prevented their departure?

Churchill held firm:

> It is not on any grounds of security but on a question of getting work done here which we have got to get done for the purpose of the Japanese war.[33]

Blackett reacted by organising a letter[34] of protest to Sir Henry Dale, co-signed by four of the 'banned eight', remonstrating about the gross interference with their freedom:

> Many Fellows of the Royal Society have loyally given their service to the Government in times of war, but they did not imagine that as a result they would be debarred from those contacts with their colleagues abroad which are the lifeblood of science.

The mess, however, was not of the Royal Society's making.

Milne distanced himself from Blackett's intemperate letter and wrote separately to Dale, assuring him that apart from the personal inconvenience

---

[33] *Hansard* (14 June 1945), Col. 1780–1781.
[34] Bernal, Blackett, Dirac, Mott and Norrish signed the letter of 14 June 1945, Box 93, Dale papers, Royal Society.

and discourtesy of cancelling the trip at a few hours' notice, he did not suspect the government of some sinister purpose. He hinted that he had an inkling of the exceptional reasons behind the government's decision:

> Publicly I accept the reason for the refusal ... The flimsiness of the reason, however, and the shortness of the notice, suggest that there were other reasons. ... I believe that there may be reasons of state which make it impolitic to disclose the true reasons for refusal.[35]

At this juncture the British, but not the Soviets, were still fighting the Japanese, and it is not inconceivable that Milne had gleaned the real reason, possibly from Darwin, whom he had known since their Cambridge days of vint and rock climbing. Only the previous October he had accompanied him and his wife to dinner at the RAS Club after Darwin had given a lecture commemorating his father Sir George Darwin.[36]

Before World War I, C.G. Darwin, along with Bohr and Rutherford, had investigated the atomic nucleus. During World War II Darwin was Director of the British Central Scientific Office in Washington and played a crucial role in developing the atomic bomb with the Americans. In fact, he had guessed the 'special circumstances'[37] before Anderson revealed them to him, namely that the atomic bomb would be ready for use in August. Until that moment Anderson and Churchill were the only people in Britain privy to this supremely sensitive information.[38] Churchill guarded it with the utmost secrecy and deliberately withheld it from the rest of the cabinet and his defence advisors.[39] Even Attlee was kept in ignorance until he became Prime Minister in July.[40] Darwin wanted to dampen the ill-will fermenting among scientists about the embargo and

---

[35]Letter from Milne to Sir Henry Dale, 18 June 1945, Box 93, Dale papers, Royal Society.
[36]Sir George Howard Darwin (1845–1912). Mathematician and astronomer. Plumian Professor at Cambridge from 1883 to 1912.
[37]Letter from C.G. Darwin to Lord Woolton, 18 June 1945, Box 93, Dale papers, Royal Society.
[38]Gowing, M. *Britain and Atomic Energy*, Oxford: Macmillan, 1964, p. 371.
[39]Ibid., p. 359.
[40]Wheeler Bennett, J.W. *John Anderson*, Oxford: Macmillan, 1962, p. 327.

tried, without success, to persuade Lord Woolton to inform the other seven scientists 'by word of mouth [of] the true reasons'.[41]

Churchill could not possibly jeopardise the use of the bomb by letting its existence inadvertently leak to the Soviets. Indeed, it could be argued that by retaining Darwin and other atomic scientists, Churchill prevented exposing them to interrogation. In the critical weeks between VE Day and VJ Day another factor compounded Churchill's dilemma. He was negotiating with Stalin over preparations for the Potsdam conference,[42] which would result in dividing Berlin into zones of occupation. To cite 'security' would be unspeakably offensive just when the Soviets were showering the British delegates with sumptuous hospitality. So, caught in the cleft of delicate circumstances, Churchill dissembled and misled the House of Commons. With hindsight of the imminent Cold War, one might conjecture whether the Soviets had an ulterior motive in their invitation, but nothing supports this view, and four years would pass before they were ready to test their own atomic bomb.

The USSR did everything in their power to impress the British delegates who were allowed to go, and they came home feeling undervalued by their own government. Loud were their praises[43] for the way the USSR integrated scientists into their society. Compared to their Western counterparts, Soviet scientists enjoyed high status within their own country with better housing, longer holidays and larger rations. The fortnight's cornucopia of meetings, excursions to museums and laboratories, including Pavlov's, a string of receptions and dinners, a banquet in Leningrad's Uritsky (Tauride) Palace, as well as concerts, ballet and opera, came with the freedom to wander unsupervised in Moscow and Leningrad. The climax was the lavish banquet in St George's Hall in the Kremlin, attended by Stalin, Molotov and other Party leaders.

On 6 August the Allies dropped the atomic bomb that demolished Hiroshima. Three days later a second bomb destroyed Nagasaki and the

---

[41] Letters from C.G. Darwin to Sir Henry Dale and Lord Woolton, 18 June 1945, Box 93, Dale papers, Royal Society.
[42] CAB 120/858, NA.
[43] *Nature* 156 (25 August 1945), p. 221; *The Times* (7 July 1945); *The Daily Mail* (10 July 1945); *News Chronicle* (10 July 1945).

Japanese surrendered. As soon as Darwin's key involvement in the atomic energy programme became public he was the hero of the hour along with John Cockcroft.

It immediately became obvious that the true reason behind the ban was security. Blackett was a prominent member of the Maud Committee which co-ordinated the work of several research groups that put Britain in the lead over America during the early stages of the war. (In the later stages of the war his advice on bombsights and radar reduced shipping losses by Nazi U-boats.) Rideal's technique for separating isotopes by centrifuge was vital in producing fissile atomic material. Whether Dirac's theoretical contributions put him in the same camp as these atomic physicists is not clear.[44]

Yet the other four, Bernal, Milne, Mott and Norrish, were not atomic physicists. Bernal evaluated the efficacy of bombing raids and, as personal assistant to Lord Mountbatten, advised on physical factors affecting the Mulberry artificial harbours for D-Day. Milne had worked exclusively on gunnery and knew nothing whatsoever about the atomic programme. Mott's involvement with nuclear physics had finished in the early 1930s; during the war he was superintendent of theory at the Armaments Research Department, Fort Halstead, concerned with conventional weapons. Norrish was an authority on incendiary devices. Not one of them was remotely associated with atomic energy. They were detained as camouflage to dilute the absence of the atomic scientists and blur their significance.

It was Milne's bad luck that he was picked to be one of the decoys, especially at this low point in his life when some relief was in order. He harboured regrets about missing such a remarkable trip and he never visited the USSR. Under Stalin's cruel dictatorship science was increasingly politicised as he insisted it confer economic benefit in accordance with Communist Party dogma. Only astronomers who belonged to the Party had survived the terrible purges of the 1930s.[45] Studies in cosmology were

---

[44]Dalitz, R.H. and Peierls, R. 'Paul Adrien Maurice Dirac' in *Bio. Mem. F. Roy. Soc.* 32 (1986), p. 154.

[45]Eremeeva, A.I. 'Political Repression and Personality: The History of Political Repression Against Soviet Astronomers' in *Journal of the History of Astronomy* 26 (1995), p. 299, and Graham, L.R. *Science in Russia and the Soviet Union*, Cambridge: Cambridge University Press, 1993, p. 221.

repressed. Since the Party demanded a Universe infinite in time and space,[46] it rejected all theories of a big bang and an expanding Universe because they incorporated the origin and evolution of the Universe. Kinematic relativity was untenable to Stalin, condemned for its reliance on the concept of divine creation.

In 1947 the Central Committee of the Communist Party orchestrated a conference, ostensibly to discuss a book[47] on Western philosophy, but actually for the purpose of strengthening the Party's position. Not a single professional astronomer or physicist was invited to take part. One of Stalin's innermost circle of brutal henchmen, Andrei Zhdanov, sarcastically castigated Milne's kinematic relativity for being unacceptable to Marxist ideology.[48]

At first sight, Soviet disapproval of Milne's ideas appears at odds with Haldane's support, but the discrepancy is explained by their different brands of communism. Although Haldane selectively endorsed scientific theories which matched his political sensibilities, he did not accept that science should be subordinated to politics, nor did he agree with Stalin's diktat that communism was a state religion. In 1950 Haldane resigned from the Communist Party. [49]

---

[46]Kragh, H. *Cosmology and Controversy*, Princeton: Princeton University Press, 1996, p. 262.
[47]Alexandrov, G.F. *The History of Westeuropean Philosophy*, London: Lawrence and Wishart, 1949.
[48]Letter from A.S. Sharov to the author, 17 May 1994.
[49]Clark, R. *J.B.S. Haldane*, London: Hodder and Stoughton, 1968, pp. 189, 185.

Chapter 17

A Race Unfinished

Posterity will put us all in our proper, humble places and will judge us chiefly for our courage in stating the truth as we see it.[1]

At the end of July 1945 Beatrice came home and Milne rejoiced that she was cheerful and affectionate. Her state of mind was fragile though, and to his distress she lapsed into melancholy silences. On 27 August she went to London to see her dentist and listen to a lunchtime concert at the National Gallery. Despite this veneer of normality, she was not in her right mind and the next day, while Milne was out of the house, she gassed herself. He had the horrifying experience of discovering her body on the kitchen floor. He wrote to Whitrow, 'She never complained and she never explained.'[2] Her death was a 'dismal mystery'[3] and she remains an enigma. He found it scarcely credible that the same tragedy had twice befallen him. Although the circumstances differed, as did his wives' temperaments and personalities, he was the common link. Had he failed them in some way? Was his absorption in his work to their detriment? G.H. Hardy held that of all the arts and sciences, mathematics is the most austere and remote.[4] A mathematician's peaks of high excitement are largely inaccessible to his wife, unless she, too, is a mathematician and can fathom their fascination. Yet professional life is but a single component among many that shape the stability and happiness of a marriage.

Family and friends rallied, yet in some quarters of the university Milne was perceived to have incurred shameful notoriety. Jean and Edwin had

---

[1]Letter from Milne to S. Chandrasekhar, 13 June 1935, EAM, MS.Eng.misc.b.427.
[2]Letter from Milne to G.J. Whitrow, 4 September 1945, GJW, private access granted by GRW, contact Imperial College Archives.
[3]Letter from Milne to W.H. McCrea, 18 September 1945, WHMcC.
[4]Hardy, G.H. 'Mathematics in Wartime' in *Eureka* 3 (January 1940), p. 8.

him to stay and the McCreas generously offered to care for the children but Milne declined as they were settled at the Children's Home. George Foster Carter's daughters helped him in practical ways and their father gave him spiritual comfort. Milne had a great regard for Foster Carter, whose wife was a Chavasse. Her father, the evangelical churchman Bishop Francis Chavasse[5] founded Oxford's St Peter's Hall (now College) and her brother Christopher[6] was its first Master. He and his twin brother, Noel,[7] twice awarded the Victoria Cross in World War I, had run in the 400 metres in the 1908 Olympics. In lengthy, consoling discussions Foster Carter strengthened Milne's conviction that faith and science are complementary and persuaded him that 'in some wonderful way the most tragic happenings have some inner purpose for our good'.[8]

Since January, Milne had lived in a twilight of uncertainty. Now he could take control and make decisions. His overriding wish was to reclaim his scattered children and he sent for my sister and me forthwith. We came home with one of Richard Southwell's daughters on the *Queen Elizabeth*, which was fitted out as a troop ship. My father greeted us warmly at the Southampton docks but unfortunately our arrival coincided with the start of the Oxford academic year. He was worn to a frazzle by the competing demands of his university commitments and getting us ration books, buying our school uniforms and putting meals on the table. He wrote, 'The struggle for existence is too great. I am suffering from nervous exhaustion, and the pain in my chest is more pertinacious.'[9] We were eager to meet Alan and Edith and we visited them in Worthing (Fig. 21), but before my father fetched them home he installed a housekeeper. That Christmas we were united but we were a family in name only, without cohesion. Like most evacuees returning from America, my sister and I were ill-prepared for Britain's grey austerity and my father, not having seen us growing up, was ill-equipped to relate to adolescent girls. All this made for a strained and joyless household. Moreover, I was totally ignorant of the turmoil he

---

[5]Francis James Chavasse (1846–1928). Principal of Wycliffe Hall, Oxford, from 1889 to 1900. Bishop of Liverpool from 1900 to 1923.

[6]Christopher Maude Chavasse (1884–1962). Bishop of Rochester from 1940 to 1960.

[7]Noel Godfrey Chavasse (1884–1917). Medical doctor.

[8]Letter from Milne to Lady Jeans, 19 October 1946, courtesy of C.V. Jeans.

[9]Letter from Milne to M.P. Dunham, 8 October 1945, EAM, MS.Eng.misc.b.429.

had suffered and he never mentioned Beatrice. His Parkinsonism created an impenetrable barrier, his glassy face devoid of emotion. He spoke only from necessity and was easily agitated which increased tensions. I had no idea that his medical condition was the reason for his lack of communication. The moments when I felt closest to him were in the evenings, after I had finished my homework, over games of piquet. He kept a running total of our scores, reminiscent of his Trinity vint.

My father realised that his reluctance to talk made him dull company and when he thanked Jan Oort for his hospitality, while a guest of the Dutch Physical Society, he apologised, 'You must have thought me often unaccountably silent during our encounters. For some reason the gift of my easy conversation has forsaken me for the time being. I hope it will return.'[10] It would not return and, perhaps to save him more worry, his doctors refrained from telling him why. His consultant was the distinguished neurologist Ritchie Russell,[11] who prescribed medicine for his palsied arm. Nevertheless, if Milne paused to consult his notes while lecturing, chalk in hand, the involuntary tremor caused a distracting rat-tat-tat on the board. The exertion of hurrying for a bus brought on tight chest pains and a gripping sensation in his arms, which he could relieve instantly with a dose of glyceryl trinitrate, used to treat angina. An annoying tingling in his legs stopped him from getting a good night's sleep.

I think he was concerned about his health and what might become of us. Heart failure had killed his own father, and in a gloomy moment my father correctly foretold that he would not see the end of food rationing in his lifetime. He was beset by worries. Besides the inflation and high taxation that hit everyone hard, he laboured under an appalling financial burden, a hurtful legacy of Beatrice's death. She left her entire estate — and it was considerable — to her daughter. Although he received not a penny, he was liable, as her husband, under 'double' taxation rules, to pay tax on income accrued from her American investments. The enormous bill exceeded his annual salary. He was frightened that he might have to use his savings, which were

---

[10]Letter from Milne to J.H. Oort, 24 September 1946, courtesy of Jet Katgert, Leiden Observatory.
[11]William Ritchie Russell (1903–1980). He is commemorated in the Russell Cairns Unit at the John Radcliffe Hospital.

modest from supporting his mother and educating Philip, and exercised the most stringent economy. Clearly it was inequitable that he should have to foot the bill from his own pocket and he lodged an objection. The protracted negotiations that ultimately reduced the bill dragged on past his death.

He was in a state of perpetual anxiety lest the housekeeper should give notice and leave him in the lurch — housekeepers were scarce because the war had opened up new opportunities for women. One housekeeper stayed for two years, another for two weeks. After a particular housekeeper took umbrage at a remark from one of us about India's forthcoming independence, he forbade us to voice opinions in her presence, which blighted mealtime conversation.

Fuel rationing was still in force and some housekeepers took advantage of their position. In the bitter winter of 1947, when Port Meadow froze to a skating rink, a profligate housekeeper used up our anthracite by stoking the boiler for a hot bath every afternoon while the rest of us were out, and the house grew stone cold. Such was the measure of sympathy for Milne that our good neighbours the Drivers[12] came to our rescue. They gave us some of their own coal ration, which we gratefully trundled home in our wheelbarrow from their house in Charlbury Road. We made sure that we kept charge of food parcels from kind Americans and reserved them for special occasions. One Boxing Day the family who came to tea were so dazzled by the confections gracing our table that they thereafter dubbed a big spread a 'Milne' or 'half a Milne'.

By the clichéd yardstick of health, happiness and prosperity, my father scored low, but he was not to be defeated and seldom referred to his troubles. Admittedly, he looked frail and elderly, 'a walking corpse',[13] as if a 'wisp of wind would blow him away'.[14] If he nodded off in church during the sermon people attributed it to old age whereas he was barely fifty. Appearances can be deceptive and he was more alert than he seemed. In an episode that went round Oxford's High Tables he confounded his

---

[12]Sir Godfrey Rolles Driver (1892–1975). Hebrew scholar.

[13]Letter from J. Collie to the author, 9 December 1991, citing his father, the physicist Carl Howard Collie (1903–1991).

[14]Conversation with Ian M. Crombie, tutor in philosophy and Fellow of Wadham College from 1947 to 1983, 31 October 1989.

Wadham colleagues when they were discussing whether or not to demolish the historic Hollywell Music Rooms owned by the college. Someone remarked that it would be a shame to destroy the oldest concert hall in Oxford, whereat Milne woke up, opened his eyes and declared with irrefutable logic, 'You can't destroy the oldest concert hall. There will always be one.'[15]

He made no concessions to his health and began the day with an invigorating cold bath before dressing in a dark suit, his gold watch chain draped across his waistcoat, with penknives, pencils, diary and a little account book in their allotted pockets. After dealing with correspondence, he set off for the Mathematical Institute, precariously mounting his bicycle and clutching his attaché case to the handlebars. He never left the house without a hat, a shabby Trilby or a surprisingly elegant summer Panama. He was home for our high tea at 5:30 pm. Evenings were for reading, often a book he was reviewing, or something solid to replenish his well-stocked mind, such as Gibbon's *Decline and Fall of the Roman Empire*. On Wednesdays he went across the road to Mrs Underhill to swap his *Spectator* for her *Punch*.

He said grace before our meals and a nightly benediction. On Sunday he recharged his soul. He loved the rhythmic cadences of the Anglican liturgy and the ceremonial of a church service, 'a form of art, of aspiration, and of worship of the mystery of things which appeal to me'.[16] We went to matins at St Andrew's or to Wadham Chapel, where we were the only family among the dons and undergraduates. Afterwards we might cycle to the Wolvercote Cemetery to put flowers on my mother's grave. My father was deeply attached to Wadham Chapel and occasionally read a lesson. One Armistice Sunday he preached a sermon on the role of a college chapel and the need for corporate worship as well as private prayer.

Wadham gave Milne adult companionship. He dined in the college on Tuesday after his colloquium and hosted lunch parties there. Every year he entertained the Halley lecturer to lunch and invited Lord Cherwell,

---

[15]Telephone conversation with the economist Sir Henry Phelps Brown (1906–1997) on 30 May 1991.

[16]Letter from Milne to S. Chandrasekhar, 22 February 1946, EAM, MS.Eng.misc.b.428.

although he didn't much care for him, with the assurance that the kitchens would provide him with vegetarian food. Milne took on the job of Wadham's librarian[17] and learned from Gerald Whitrow's wife, Magda, how to catalogue acquisitions. He regulated the use of the library, which in those days was above the kitchens to prevent the books getting damp, as stipulated by Dorothy Wadham.

He contrived to travel abroad to conferences, usually only once a year, while our housekeeper took her holiday and we four children went to relatives in Edinburgh. At a conference on theoretical science[18] in Brussels, led by Dominicans, which gave it a Catholic flavour, Milne found himself in the company of Herbert Dingle. They were the only guests from Britain, the rest being from Belgium, Holland, France and Switzerland. Given their long history of disagreement, it was ironic that they should be thrown together. Yet Dingle in person, dapper and docile, was the acme of courtesy, quite unlike Dingle in print with his rapier-like attacks and devastating satire. Over a glass of beer at a café Milne enjoyed getting to know a fresh group of physicists and philosophers, especially Jean-Louis Destouches,[19] who was keenly interested in Milne's approach to time.

In Paris an assistant at the Gauthier-Villars Bookshop amused Milne with his technique for avoiding bureaucratic regulations. After buying a copy of Fourier's *La Théorie Analytique de la Chaleur*, Milne asked him to send it to Oxford. The assistant explained that this would necessitate obtaining an import licence and so instead he intended to put the address of Milne's hotel on the package, then cross it out and add the Oxford address. The book arrived safely.

Milne was accustomed to living in a hurry and he did not slacken his pace. Lectures and meetings took him to Hull, Manchester, London and Cambridge. At one point he went to Cambridge twice in a single week, to set up the Eddington Memorial Lecture and to elect Littlewood's successor to the Rouse Ball Chair.

Far from cutting back on his workload, he added administrative responsibilities. Oxford University was sufficiently open-minded to accept

---

[17] It carried an honorarium of £25.

[18] *Bulletin de l'Academie International de Sciences Theoretiques* (1948–1950), pp. 7–25.

[19] Jean-Louis Destouches (1909–1980). Quantum theory physicist. He taught at the University of Paris.

money from industry and welcomed the offer by Imperial Chemical Industries (ICI) to fund twelve fellowships, each of three years tenure with a handsome stipend of £600 per annum. From its inception in 1944, Milne chaired the committee overseeing the fellowships and laid down terms of reference. He insisted that the Fellows do some teaching because otherwise 'research work tends to become desiccated'.[20] The chairman of ICI, Lord McGowan,[21] was pleased with the arrangements and renewed the scheme for another seven years. Similarly, Milne set up and chaired a benefaction from the Pressed Steel Company (later part of British Leyland), but the company disapproved of a part-time chairman and demanded the university provide a full-time executive. In 1948 Milne became chairman of Oxford's Board of Physical Sciences.

The drafting of rules suited Milne's quest for orderliness and he was part of a small committee that overhauled procedures at the RAS. Reforms that lasted some forty years included a provision that ensured a healthy turnover of council members to prevent stagnation.

Yet administration was merely a sideshow. Milne's defence work never ceased and the Royal Artillery Institution at Woolwich made him an honorary member. He continued to sit on the Scientific Advisory Council, which underwent re-organisation in 1946. In its new structure,[22] he chaired the renamed Ballistics Research Committee and served on the Armaments Research Board. Following a further shake-up in 1948, he chaired its Liquid Filled Shell Stability Panel and joined the Weapons Research Committee. In all, his ballistics work spanned thirty-four years, starting from the Darwin–Hill Mirror Position Finder. Remarkably, despite the technical advances during this period, the device did not become obsolete because it took half the time of an acoustical method to locate a high elevation shell burst.[23]

Milne had lost nothing of his appetite for new ideas and passed one afternoon with a young chemist, Michael Dewar,[24] who was impressed by

---

[20]Milne, E.A. 'ICI Fellowships' in *Oxford* 9 (1945), pp. 22.
[21]Harry Duncan McGowan, Baron McGowan (1874–1961). Chairman of ICI from 1940 to 1960.
[22]The council was split into four boards, on chemistry, munitions, armament research and biology.
[23] Minutes of Weapons Research Committee, WO195/11191, NA.
[24]Michael James Steuart Dewar (1918–1997). He held chairs at Queen Mary College, London, and the Universities of Texas and Florida.

Milne's willingness to spend so much time with a stranger. Dewar thought that variations in the velocity of light could lead to a new theory of relativity and consulted his cousin, the Aristotelian philosopher Sir David Ross,[25] who referred him to Milne. Intuitively Milne felt that Dewar's proposal was faulty but could not nail the error and eventually spluttered, 'This is clearly nonsense. Anyone can see that it is nonsense, but I can't see why it is nonsense. You must publish it so that other people can find what is wrong with it.'[26] Dewar wrote up his train of thought in a paper[27] and Milne saw that it got published, though curiously Dewar omitted it from his list of lifetime publications.

Students noticed Milne's simplicity and his sense of what was proper. A budding physicist who raised a point about conservation theorems was disarmed by Milne's frankness when he replied with great intensity, 'That is very interesting but I have no time to think about it right now. I am trying to conserve the momentum of the universe.'[28] This conjured up an image of Milne struggling single-handedly to conserve momentum as if, should he fail, the Universe would collapse.

Milne relished historical support for kinematic relativity and took childish glee in learning that the cosmological principle was enunciated as early as the sixteenth century by the wandering Italian philosopher Giordano Bruno.[29] When he was in England Bruno seemed to support the principle by affirming 'that the eye being placed in any part of the Universe the appearance would be still all one as unto us here'.[30] Milne revelled in being in Bruno's company, though David Kendall, who unearthed the quotation, was not sure whether this was despite or because

---

[25]Sir (William) David Ross (1877–1971). Provost of Oriel College from 1929 to 1947.

[26]Dewar, M.J.S. *A Semiempirical Life*, Columbus: American Chemical Society, 1992, pp. 33–34.

[27]Dewar, M.J.S. 'The Interpretation of Light and its Bearing on Cosmology' in *The Philosophical Magazine* 38 (1947), pp. 88–94.

[28]Letter from R. Wilson to the author, 6 March 1989.

[29]Giordano Bruno (1548–1600). He was excommunicated by Catholics, Calvinists and Lutherans alike.

[30]Letter from Sir William Lower, dated 21 June 1610, to the Oxford astronomer Sir Thomas Harriot, tutor and companion to Walter Raleigh. Quoted in a letter from D.G. Kendall to Milne, 26 January 1950, in the author's possession.

Bruno was burnt at the stake for heresy. Perhaps Bruno inspired Milne to write 'Scientific Heresies'[31] in which he deplored those who blindly follow dogma. He urged them to have the courage to think for themselves: 'The essence of scientific freedom is the right to come to conclusions differing from those of the majority.' For his part, he averaged about two original papers a year at this time. The most noteworthy was one in which he calculated the shape of the arms of a spiral nebula and showed that it tallied with observation. During these years he was elected to the American Academy of Arts and Sciences and the Calcutta Mathematical Society, and the Cambridge Philosophical Society awarded him its Hopkins Prize.

That he produced fewer original papers did not mean his pen was idle and when in full spate he 'could have kept two pens busy'.[32] Besides four books, he wrote for *The Oxford Magazine*, *The Dictionary of National Biography*[33] and contributed a chapter to *Albert Einstein*,[34] a volume that commemorated his seventieth birthday. Milne's special forte was obituaries. The loss of his father, brother and two wives brought a compassionate poignancy to his portrayal of cherished colleagues. *The Times* filed his 'stock' obituaries for future use. After Milne's own death, *The Times* simply 'topped' and 'tailed' what he had previously prepared when F.J.M. Stratton died in 1960 and Plaskett in 1980.

Writing the obituary or biographical memoir of a Fellow of the Royal Society, which stands in perpetuity as the definitive assessment of his/her scientific oeuvre, is a major undertaking. Milne may hold a record in writing four and contributing to a fifth as joint author. And he completed them promptly. He wrote three in difficult circumstances and without autobiographical notes: A.E.H. Love's during the Blitz, H.F. Newall's during the doodlebug raids and R.H. Fowler's while homeless in Hessle.[35]

---

[31] A copy is in the author's possession.

[32] Cowling, T.G. 'Astrology, Religion and Science' in Bondi, H and Weston-Smith, M. (eds) *The Universe Unfolding*, Oxford: Clarendon Press, 1998, p. 69.

[33] Milne wrote entries for R.H. Fowler, A.E.H. Love and H.F. Newall.

[34] Schilpp, P.A. (ed.) *Albert Einstein*, Evanston: The Library of Living Philosophers, Inc., 1949, pp. 409–435.

[35] The other two were on J.H. Jeans and, with Francis Puryer White (1893–1969), on H.W. Richmond.

Milne missed Fowler acutely and his sensitive obituary reached a literary pinnacle. Hill judged it the finest bit of sympathetic writing he had ever read and hoped Milne would write his own, but Milne pre-deceased him.

Milne was commissioning editor for a series of inexpensive little books for the lay reader about mathematics and the physical sciences, published by Hutchinson's University Library. To prospective authors he took pains to define the standard and length required: 60,000 words, which allows for 'the cream but not the milk'.[36] His authors did not let him down and several titles were reprinted or went into further editions. The first to appear, later translated into Turkish, was *Geometry* by Henry Forder, who had a knack for catchy maxims; for example, he summed up Archimedes' axiom with the words, 'You will always reach home if you walk long enough.'[37] A spectacular triumph was Harold Davenport's *The Higher Arithmetic*, a minor classic now in its eighth edition. When Milne asked Davenport to write the book, he caught him at an awkward moment, as his hands were full taking over the Mathematics Department at University College London. Nonetheless, Davenport felt compelled to oblige, 'If Professor Milne has asked me, I shall have to do it.'[38]

The post-war paper shortage delayed the publication of *Vectorial Mechanics*[39] but it finally appeared in 1948 and Milne stopped lecturing on the subject. The book became a standard text, reprinted in 1957, and sold well for years. Its crusading message is the usefulness of the vector product in dealing with spinning tops, rolling spheres and the like. (The use of the symbol $\wedge$ which avoids the ambiguity of $\times$ is largely due to Milne, Chapman and Hartree.) Milne's approach influenced the writer of *Vectorial Astronomy*,[40] and inspired the cosmologist David Matravers[41] to call his inaugural lecture at Portsmouth University 'Of Milne and Mathematics'.

Six weeks later Milne's *Kinematic Relativity* came out. In it he explored the duality of the Universe through two time scales and wrote up his

---

[36] Letter from Milne to W.H. McCrea, 21 January 1946, WHMcC.

[37] *Bull. London Math. Soc.* 17 (1985), p. 164.

[38] Conversation with A. Davenport on 17 October 1990.

[39] Milne, E.A. *Vectorial Mechanics*, London: Methuen and Co. Ltd, 1948.

[40] Murray, C.A. *Vectorial Astronomy*, Bristol: Adam Hilger, 1983.

[41] David Matravers. Emeritus Professor at the Institute of Cosmology and Gravitation, Portsmouth.

reconstruction of the laws of dynamics, gravitation and electromagnetism. He left his Christianity to the last paragraph with a plea that 'it requires a more powerful God to create an infinite Universe than a finite Universe.'[42]

The book, described as an 'intellectual adventure',[43] sparked a flush of interest. Oxford University Press considered republishing, which gave Milne the chance to correct a mistake relating to the photon. His attitude to errors was ambivalent. He readily acknowledged an error when it was a lapse in mathematical logic but if it hinged on physics he was slow to give way. He was determined to find a way round the mistake, which was brought to his attention by a Cambridge graduate student, Allan Curtis,[44] who was preparing a doctoral thesis on cosmology. Curtis was also trying for a fellowship at St John's, and the college asked Milne to assess his work. He recommended Curtis, with the consequence that because the fellowship secured Curtis's future he abandoned his thesis. Milne resolved the error Curtis had pointed out by assuming that Planck's constant varies with the age of the Universe and revised the relevant chapter.[45] The Press, however, changed its mind about republishing and Milne fitted his correction[46] into another book.

Before Milne had finished writing *Kinematic Relativity* he unwisely signed a contract with Cambridge University Press for a biography of Sir James Jeans. Hutchinson had planned a book on Jeans to be jointly written by Sydney Chapman, Harold Jeffreys and Milne, but Jeans's second wife, the distinguished organist Susi Jeans,[47] insisted that S.C. Roberts,[48] a friend of Jeans's who had supervised his popular writings, be in charge. Milne was not Roberts's first choice. The physicist and *Brains Trust*

---

[42]Milne, E.A. *Kinematic Relativity*, Oxford: Oxford University Press, 1948, p. 233.
[43]*Nature* 163 (18 June 1949), p. 931.
[44]Allan Raymond Curtis (1922–2008). Mathematician. His major appointments were at Harwell and the UK Atomic Energy Authority.
[45]Manuscript in the author's possession.
[46]Milne, E.A. *Modern Cosmology and the Christian Idea of God*, Oxford: Oxford University Press, 1952, p .134.
[47]Susi Jeans, Lady Jeans, (1911–1993). Musicologist and teacher.
[48]Sir Sydney Castle Roberts (1887–1966). Author and publisher. Secretary of Cambridge University Press from 1922 to 1948. Master of Pembroke College, Cambridge, from 1948 to 1958.

panellist, Edward da Costa Andrade,[49] had refused on the grounds that he was neither an astronomer nor a cosmologist. Milne had the opposite handicap; he knew Jeans only through astronomy and cosmology.

Few could surpass Milne's admiration for Jeans. From his school days, Milne had lapped up Jeans's books and had reviewed some of them. Yet, although they had been in one another's company on countless occasions, they were not close. Indeed, Jeans had few friends. In his prime he retreated to Cleveland Lodge, Dorking, with his first wife, Charlotte (née Tiffany, the jewellers), where he built two organs and devoted himself to music and science.

Besides the Royal Society obituary on Jeans, Milne had supplied lesser notices upon his death,[50] one at the request of George Trevelyan. For the book, Milne needed to fathom the man behind the scientist and arranged to talk to Lady Jeans. A crisis over housekeepers postponed his visit to Cleveland Lodge but when he did stay the night she was reticent. Even appeals to the readers of *Nature* and *The Times Literary Supplement* produced nothing that shed light on Jeans's personality. It is pertinent that Milne underlined in his copy of Boswell's *Life of Johnson* a comment about an author and his manuscript: 'Not that it is poorly written, but that he had poor materials.'[51] Milne's draft chapters upset Lady Jeans because she felt Milne had not conveyed Jeans's air of distinction nor done justice to his musical talents. He had created the impression that Jeans was aloof whereas she knew his love of laughter and dry humour. In the main, the book is a critique of Jeans's scientific legacy yet even in this Milne fell short. He had not noticed that in the last edition of *The Universe Around Us* Jeans swung round to Eddington's views on stellar structure.[52] Milne died before the book was published and although S.C. Roberts added a personal memoir, the book is unsatisfactory. Some people hold the view

---

[49]Edward Neville da Costa Andrade (1887–1971). Physicist. Professor at University College London.

[50]*Nature* 158 (19 October 1946), pp. 542–543; *Proc. of the Physical Soc.* 59 (1946), pp. 503–506; *Journal of the Lond. Math. Soc.* 21 (1946), pp. 310–320; and by request *Cambridge Review* (2 November 1946), pp. 76–78.

[51]Boswell, J. *Life of Johnson*, London: Henry Frowde, 1904, p. 449.

[52]W.H. McCrea speaking on the BBC Third Programme, 15 January 1954, *I.A.J.* 4 (1957), pp. 23–28, courtesy of Leon Mestel.

that a biography should not follow hard on the heels of the subject's death, perhaps illustrated by this case.

Milne's fascination with time drew him to realms outside science. He once wrote, 'If astronomy does not manufacture philosophers, no science can',[53] and at the British Royal Institute of Philosophy he spoke on the role of timekeeping. He served on the editorial board of *Philosophy* and reviewed books, including one on Aristotle and other ancient philosophers. For the *Hibbert Journal*, which straddled religion and philosophy, he reviewed[54] Bertrand Russell's last major work, *Human Knowledge*, and was not alone in finding it a disappointment.

While praising the elegance and clarity of Max Born's *Natural Philosophy of Cause and Chance*, Milne grabbed the opportunity to refute his criticisms of Milne's methods. Born had devoted an entire lecture to denigrating Milne's deductive approach and ridiculing his use of light signals as 'weird inventions'.[55] Milne retorted that Born's dismissal of measuring time by light signals did 'not justify his burying his head in the sand like an ostrich'.[56]

June was a busy month in Milne's social calendar with events like the Open Day of the Royal Greenwich Observatory — Milne served on its board. Oxford had a round of summer garden parties and in 1950 my father decided that at the age of seventeen I should accompany him. At the Encaenia garden party in Christ Church, tucking into delicious strawberries and cream, I was surprised by how vivacious my father was and how cordially people engaged with him. The same was true of the Royal Society's Ladies' Soirée in Burlington House. Awed by the full evening dress occasion, and clearly the youngest present, I saw my father's face break into a radiant smile as he greeted his hero[57] Bishop Barnes, resplendent in silk breeches and gaiters. These occasions gave me a glimpse of

---

[53]Milne's review of Eddington's *The Nature of the Physical World* in *The Listener* (10 April 1929), p. 466.

[54]*Hibbert Journal* 47 (April 1949), pp. 297–299.

[55]Born, M. *Experiment and Theory in Physics*, Cambridge: Cambridge University Press, 1943, p. 41. Delivered to the Durham Philosophical Society at Newcastle on 21 May 1945.

[56]*Philosophy* 24 (1949), p. 371.

[57]Milne admitted to 'a little hero worshipping' in his review of Barnes's *Scientific Theory and Religion* in *The Oxford Magazine* (1 June 1933), p. 730.

someone warm and outgoing, not at all like the tense, silent person I knew at home.

That June held a nostalgic treat. As if fate decreed he should have a final visit to the springboard of his career, Milne attended gun trials at HMS *Excellent*. From the minute he rattled across the bridge to Whale Island, he savoured every moment of his overnight stay. He dined in the Mess, a band playing and the silver agleam; he boarded a pinnace to look at the gun turrets on a warship moored in the harbour; and he inspected equipment at Eastney, where the wind had blown him to bits at the grid and gantry. He climbed up the steep, narrow staircase to the eaves where he had toiled with Fowler and Richmond. Overflowing with sentimental memories, he immediately gave an account of his visit to Hill.[58]

It is evident that by 1949 Milne had no doubts about the parlous state of his health for when he proposed Geoffrey Walker for fellowship of the Royal Society, he felt obliged to reveal the names of his seconders, 'in case anything happens to me'.[59] In September 1950, disregarding medical advice to reduce his activities, he set off for an RAS conference in Dublin, intent on giving his latest paper[60] which linked gravitation and magnetism. He took a train to Birmingham, changed stations there and at Liverpool he met McCrea for the night ferry across the Irish Sea. They were staying at the same hotel, and McCrea, who knew about Milne's weak heart, thoughtfully carried his suitcase upstairs for him. With a premonition of death, Milne told him that his affairs were in order. That afternoon Milne sent postcards to us children. At an evening reception in the fine eighteenth-century rooms of the Royal Irish Academy, he was, 'as always, bringing his juniors into the animated groups debating this point or that in the scheduled programme'.[61]

The next morning, Thursday 21 September, he left his hotel in d'Olier Street to go to a discussion on solar physics. As he was walking briskly along Dawson Street, he suddenly lurched forward and collapsed in the doorway of a bookshop. A few yards behind him was Ernest Walton, who

---

[58]Letter from Milne to A.V. Hill, 10 June 1950, AVHL I.

[59]Letter from Milne to A.G. Walker, 17 September 1949, AGW.

[60]Published posthumously by G. Whitrow. *Mon. Not. R. Ast. Soc.* 110 (1950), pp. 266–274.

[61]Letter from A. Hunter to P.S. Milne, 21 September 1950, in the author's possession.

had famously split the atom with Cockcroft. He noticed a small crowd gather, then a policeman appeared and an ambulance took Milne to hospital, where he was pronounced dead from heart failure. He was fifty-four. A notebook in his pocket revealed his identity and a doctor telephoned Alan Hunter, the secretary of the RAS, just as Milne's solar work was being discussed. Alan Hunter telephoned Wadham and Jack Thompson telephoned my headmistress. When she broke the news to me an enormous sense of relief tempered the shock.

The Oxford term had not begun and many people were away for the funeral in Wadham Chapel, but even so chairs had to be put up in the antechapel. An honorary Fellow of Wadham, the Right Reverend Campbell Hone[62] led the service, assisted by the college chaplain and George Foster Carter. My father was buried, as he wished, in my mother's grave at Wolvercote Cemetery.

Milne had left on his desk the manuscript of ten lectures, titled 'Modern Cosmology and the Christian Idea of God', that he was to give in Birmingham. The Edward Cadbury Lectures are normally the province of theologians since the brief is to consider the relation of Christianity to civilisation. That Milne was chosen was most likely due to the influence of Martin Johnson, who was a prominent member of the university. Milne's theme was the partnership between Christianity and the laws of nature interpreted through kinematic relativity.

Gerald Whitrow, Milne's literary executor who dealt with the Jeans biography, sent the manuscript to Oxford University Press, as previously planned. Trouble loomed. The Press was aware that many scientists disapproved of Milne's unconventional views and put the manuscript out to referees. Despite their mixed verdicts, the Press was willing to go ahead until the eminent chemist Cyril Hinshelwood,[63] a delegate of the Press, intervened. He insisted that Dingle be consulted. Readily acknowledging his prejudice, Dingle set out his reservations in copious detail.

---

[62]Campbell Richard Hone (1873–1967). Bishop of Pontefract from 1931 to 1938. Bishop of Wakefield from 1938 to 1945.
[63]Sir Cyril Norman Hinshelwood (1897–1967). Chemist, linguist, classicist. Professor at Oxford from 1937 to 1964. President of the Royal Society from 1955 to 1960. President of the Classical Association. Nobel Laureate 1956.

Notwithstanding Dingle's disparagement, the Press decided to publish because it would be Milne's last book. Dingle used his preview of the manuscript to ask Whitrow to alter a derogatory phrase Milne had written about him that he construed as libellous. Wisely, Whitrow deleted the whole sentence.

The book was Milne's 'swan song',[64] tactfully labelled 'provocative' and 'challenging' by reviewers.[65] That theologians were more sympathetic, notably in a five-page article in the *Church Quarterly Review*,[66] would have greatly pleased him. This was recognition on a higher plane.

Milne's obsession with time and kinematic relativity moulded the background to future cosmology. When Whitrow gave a radio broadcast[67] about Milne the steady-state theory of continuous creation was in vogue. It prevailed until 1963 when Arno Penzias[68] and Robert Wilson,[69] employed by the Bell Laboratory in New Jersey, chanced on unexplained, extraneous noise that turned out to be the cosmic microwave background. This ubiquitous relic of radiation from a cataclysmic explosion is compelling evidence for an expanding Universe. Big bang theories, chiefly Georges Lemâitre's primordial theory, regained the ascendant. Much more recently came the discovery that the Universe is 'lumpy', contains mysterious dark matter, and that its expansion is accelerating. 'Flat' space, which Milne favoured, has not been ruled out.

Milne never enjoyed the fruits of seniority, a *Festschrift* or a special birthday celebration, but his achievements in meteorology, ballistics, astrophysics and cosmology exceed those of many who live longer. He benefited from early recognition, an accelerated career and the patronage of influential people, and he never applied for a job. A crater on the Moon and a Minor Planet #11767 carry his name.

---

[64] Review by Zdenek Kopal (1914–1933) in the *Manchester Guardian* (30 March 1952).
[65] *The Observatory* 73 (1953), p. 30.
[66] *Church Quarterly Review* 153 (July–September 1952), pp. 355–360.
[67] The Third Programme on 18 December 1950 at 9:40 pm.
[68] Arno Allan Penzias (b. 1933). German-born physicist. He joined Bell Telephones in 1961. Nobel Laureate 1978.
[69] Robert Woodrow Wilson (b. 1939). American physicist. Appointed Head of Radio Physics Research at Bell Telephones in 1976. Nobel Laureate 1978.

The constants in his life, his ingrained sense of duty, his gritty tenacity and his Christian faith, gave him the strength to rise above adversity. The late John Maddox,[70] editor of *Nature*, who attended Milne's Oxford colloquia, described him as the 'nicest of men'.[71] Distrustful of the orthodox, quixotic, unusually energetic, often out of step with the majority, Milne was conspicuous in thinking for himself. He was obstinate but not presumptuous, outspoken but not flamboyant. His own words make a fitting epitaph: 'The man of science is necessarily a rebel, a prophet rather than a priest, a man who is always finding himself in opposition to the hierarchy.'[72]

[70]Sir John Royden Maddox (1925–2009). Scientific correspondent on the *Manchester Guardian* from 1955 to 1964. Editor of *Nature* from 1966 to 1973 and from 1980 to 1995.
[71]*Nature* (12 May 1988), p. 111.
[72]From 'Scientific Heresies' p. 11, in the author's possession.

Epilogue

After my father's death we children left our house in Northmoor Road, and although we were freed from our restricted existence we missed our friends in the neighbourhood. My sister Edith, aged seven, went to her Aunt Izzy and Uncle Jim in New Jersey and grew up with their daughters on their small holding. She learned to ride, became a proficient show jumper and set up her own stables. She died in a road accident at the age of twenty-four.

Eleanor, Alan and I spent our holidays with our Aunt Jean and Uncle Edwin in Hampshire, while continuing our education in term time — Eleanor at university, Alan at boarding school. I was the lucky one. To enable me to finish my sixth form studies at the Oxford High School, Theo and Hilda Chaundy, who were the soul of kindness, welcomed me into their lively household.

Eleanor married and moved to Australia, where her husband taught at Geelong Grammar School and she brought up her family. She died in 1986. Alan read economics at Trinity College, Cambridge, travelled extensively, studied at the School of Oriental and African Studies, taught history, settled in Tasmania and took up a new career in computing before retirement.

I knew of my father's international reputation and in my gap year, before going up to Cambridge, I had a summer job at the Harvard Observatory. Many years passed before I had any inkling of my father's troubles because he never spoke about them. Gathering and digesting material for this book compelled me to revise my view of him. To be a single parent with four children while working to full capacity is no easy task, and I now think he did his best for us in circumstances that would have defeated a lesser man.

Index